P9-ARK-102

APPLIED RELIABILITY

APPLIED RELIABILITY

Paul A. Tobias, Ph.D.
Senior Technical Staff Member
and Manager of Field Performance
and Statistical Support
IBM Corporation
Hopewell Junction, New York

David C. Trindade, Ph.D.
Corporate Director of Reliability
Advanced Micro Devices, Inc.
Sunnyvale, California

VNR VAN NOSTRAND REINHOLD COMPANY
New York

Library of Congress Catalog Card Number: 85–20220
ISBN: 0–442–28310–5

Manufactured in the United States of America

Published by Van Nostrand Reinhold Company Inc.
115 Fifth Avenue
New York, New York 10003

Van Nostrand Reinhold Company Limited
Molly Millars Lane
Wokingham, Berkshire RG11 2PY, England

Van Nostrand Reinhold
480 Latrobe Street
Melbourne, Victoria 3000, Australia

Macmillan of Canada
Division of Gage Publishing Limited
164 Commander Boulevard
Agincourt, Ontario M1S 3C7, Canada

15 14 13 12 11 10 9 8 7 6 5 4 3 2

Library of Congress Cataloging-in-Publication Data

Tobias, Paul A.
Applied reliability.

Includes index.
1. Reliability (Engineering) 2. Quality control—Statistical methods. I. Trindade, David. II. Title.
TA169.T63 1986 620′.00452 85–20220
ISBN 0–442–28310–5

Preface

The purpose of this book is to provide the working engineer, statistician, or scientist with practical tools and techniques for solving today's applied problems in reliability.

Quality and reliability have become strategic variables on a par with price and performance. The average consumer consults tables that compare repair records before selecting an automobile; corporations demand ever more stringent guarantees of defect-free operation when purchasing data-processing equipment. Those businesses that emphasize quality and reliability as part of their normal manufacturing procedures are the ones that will be able to compete in today's marketplace; those that regard quality assurance as a set of final inspection screens and reliability as a warranty pricing concern will find themselves dwindling away.

The importance of quality and reliability is no longer "new news" or controversial. Numerous excellent books have described how to make quality a way of life. Statistical consultants offer many courses on the mathematics and management aspects of quality, and several large corporations have established their own internal quality schools or institutes. The tools and techniques of quality control are well known and widely practiced.

The analysis and control of product reliability is not as well understood. Requiring systems to work, not only the first time, but for many hours or months or years thereafter, makes the testing and product assurance role much more difficult. There are only a few books and courses available to teach an engineer how to run the experiments and make the decisions that are required by management. Most textbooks on reliability are theoretical in nature on the one hand or comprehensive reference works on the other, neither of which fully serves the needs of the reliability engineer or statistician who has to answer reliability concerns on a daily basis.

This text is aimed primarily at those individuals who have responsibility for the design or evaluation of the reliability aspects of components or hardware systems. Statisticians desiring an introduction to the definitions, distributions, techniques, and models currently used to evaluate reliability will also benefit.

Much of this book evolved from lecture notes written by the authors for the purpose of teaching reliability and quality concepts to managers and engineers within an intensive one-week course at various IBM facilities throughout the world. The notes were compiled because no available text adequately covered all the techniques and procedures actually in use to evaluate advanced technology reliability. The selection of material was dictated by what was needed; the style of presentation was dictated by what worked.

The material varies considerably in scope and level of difficulty. Chapter 1 covers elementary descriptive statistics, whereas Chapter 8 includes models for general reliability algorithms and burn-in benefits. Chapter 6 describes how to fit a line through points, whereas Chapters 4 and 5 tell the reader why it might be beneficial to buy state-of-the-art programs for maximum likelihood estimation. Chapter 9 contains the theory of acceptance sampling plans, as well as a wide collection of charts and nomographs for choosing sampling plans and acceptance numbers. Most books on reliability do not include the quality-control procedures described in Chapter 9, but since these are often used to control reliability, they deserve a detailed description.

The reliability analyst needs to combine standard statistical methods with advanced state-of-the-art techniques, on a day-to-day basis. To do so requires being familiar with a collection of quick graphical methods and knowing their strengths and weaknesses. When extremely important decisions based on reliability data analysis must be made, the analyst should know what advanced computer programs are available for purchase. An understanding of life distributions and acceleration models and a collection of proven statistical data analysis tools are essential.

The best way to meet these needs is to illustrate the definitions, theories, and models with applied numerical problems that make use of actual or simulated data. There are sixty formal examples of this type throughout the text, as well as many informal ones. These examples illustrate many of the typical problems a reliability analyst must handle. Students are encouraged to work out as many of them on their own as possible and then check out their work with the solutions in the text. That way they can be sure of understanding the methods well enough to apply them to real data. Over ninety graphs, charts, and tables are also included to supplement the other material.

Mathematical reliability theory, especially in the areas of data analysis and modeling stress acceleration, is a rapidly evolving discipline. As time goes on, new methods will replace some of those described in this book. As of now, however, we present them as a collection of well-tested techniques that have proven successful in evaluating and predicting reliability.

PAUL A. TOBIAS
DAVID C. TRINDADE

Contents

APPLIED RELIABILITY

1
Basic Descriptive Statistics

One of the most useful skills a reliability specialist can develop is the ability to convert a mass (mess?) of data into a form suitable for meaningful analysis. Raw numbers by themselves are not useful; what is needed is a distillation of the data into information.

In this chapter we discuss several important concepts and techniques from the field of descriptive statistics. These methods will be used to extract a relevant summary from collected data. The goal is to describe and understand the random variability that exists in all measurements of real world phenomena and experimental data.

The topics we shall cover include: populations and samples; frequency functions, histograms, and cumulative frequency functions; the population cumulative distribution function (CDF) and probability density function (PDF); elementary probability concepts; random variables, population parameters, and sample estimates; theoretical population shape models and data simulation.

POPULATIONS AND SAMPLES

Statistics is concerned with variability, and it is a fact of nature that variation exists. No matter how carefully a process is run, an experiment is executed, or a measurement is taken, there will be differences in repeatability due to the inability of any individual or system to control completely all possible influences. If the variability is excessive, the study or process is described as lacking control. If, on the other hand, the variability appears reasonable, we accept it and continue to operate. How do we visualize variability in order to understand if we have a controlled situation?

Consider the following example.

EXAMPLE 1.1 AUTOMOBILE FUSE DATA

A manufacturer of automobile fuses produces lots containing 100,000 fuses rated at 5 A. Thus, the fuses are supposed to open in a circuit if the current through the fuse exceeds 5 A. Since a fuse protects other elements from possibly damaging electrical overload, it is very important that fuses function properly. How can the manufacturer assure himself that the fuses do indeed operate correctly and that there is no excessive variability?

Obviously he cannot test all fuses to the rated limit since that act would destroy the product he wishes to sell. However, he can sample a small quantity of fuses (say, 100 or 200) and test them to destruction to measure the opening point of each fuse. From the sample data, he could then infer what the behavior of the entire group would be if all fuses were tested.

In statistical terms, the entire set or collection of measurements of interest (e.g., the blowing values of all fuses) define a *population.*

A population is the entire set or collection of measurements of interest.

Note that a population may be finite as in the case of the fuses or it may be infinite as occurs in the situation of a manufacturing process where the population could be all product that has been or will be produced in a fabricating area.

The *sample* (e.g., the 100 or 200 fuses tested to destruction) is a subset of data taken from the population.

A sample is a subset of data from the population.

The objective in taking a sample is to make inference about the population.

Note that data may exist in one of two forms. In *variables* data, the actual measurement of interest is taken. In *attribute* data, the results exist in one of two categories: either pass–fail, go–no go, in spec–out of spec, and so on. Both types of data will be treated in this text.

In the fuse data example, we record variables data but we could also transform the same results into attribute data by stating whether a fuse opened before or after the 5 A rating. Similarly, in reliability work one can measure the actual failure time of an item (variables data) or record the number of items failing before a fixed time (attribute data). Both types of data occur frequently in reliability studies.

Later we will discuss such topics as choosing a sample size, drawing a sample randomly, and the "confidence" in the data from a sample. For now, however, let us assume that the sample has been properly drawn and consider what to do with the data in order to present an informative picture.

HISTOGRAMS AND FREQUENCY FUNCTIONS

In stating that a sample has been *randomly drawn* we imply that each measurement or data point in the population has an equal chance or probability of being selected for the sample. If this requirement is not fulfilled, the sample may be "biased" and correct inference about the population might not be possible.

What information does the manufacturer expect to obtain from the sample measurements of 100 fuses? First, the data should cluster about the rated value of 5 A. Second, the spread in the data (variability) should not be large, because the manufacturer realizes that serious problems could result for users of the fuses if some blow at too high a value. Similarly, fuses opening at too low a level could cause needless repairs or generate unnecessary concerns.

The reliability specialist randomly samples 100 fuses and records the data shown in Table 1.1. It is easy to determine the high and low values from the sample data and see that the measurements cluster roughly about the number 5. Yet, there is still difficulty in grasping the full significance of this set of data.

Table 1.1. Sample Data on 100 Fuses.

4.64	4.95	5.25	5.21	4.90	4.67	4.97	4.92	4.87	5.11
4.98	4.93	4.72	5.07	4.80	4.98	4.66	4.43	4.78	4.53
4.73	5.37	4.81	5.19	4.77	4.79	5.08	5.07	4.65	5.39
5.21	5.11	5.15	5.28	5.20	4.73	5.32	4.79	5.10	4.94
5.06	4.69	5.14	4.83	4.78	4.72	5.21	5.02	4.89	5.19
5.04	5.04	4.78	4.96	4.94	5.24	5.22	5.00	4.60	4.88
5.03	5.05	4.94	5.02	4.43	4.91	4.84	4.75	4.88	4.79
5.46	5.12	5.12	4.85	5.05	5.26	5.01	4.64	4.86	4.73
5.01	4.94	5.02	5.16	4.88	5.10	4.80	5.10	5.20	5.11
4.77	4.58	5.18	5.03	5.10	4.67	5.21	4.73	4.88	4.80

Let us try the following procedure:

1. Find the *range* of the data by subtracting the lowest from the highest value. For this set, the range is $5.46 - 4.43 = 1.03$.

2. Divide the range into 10 or so equally spaced intervals such that readings are uniquely classified into each cell. Here, the cell width is 1.03/10 = 0.10, and we choose the starting point to be 4.395, a convenient value below the minimum of the data and carried out one digit more precise than the data to avoid any confusion in assigning readings to individual cells.
3. Increment the starting point by multiples of the cell width until the maximum value is exceeded. Thus, since the maximum value is 5.46, we generate the numbers 4.395, 4.495, 4.595, 4.695, 4.795, 4.895, 4.995, 5.095, 5.195, 5.295, 5.395, and 5.495. These values will represent the end points or boundaries of each cell, effectively dividing the range of the data into equally spaced class intervals covering all the data points.
4. Construct a *frequency table* as shown in Table 1.2 which gives the number of times a measurement falls inside a class interval.
5. Make a graphical representation of the data by sketching vertical bars centered at the midpoints of the class cells with bar heights proportionate to the number of values falling in that class. This graphical representation shown in Figure 1.1 is called a *histogram.*

A histogram is a graphical representation in bar chart form of a frequency table or frequency distribution.

Table 1.2. Frequency Table of Fuse Data.

CELL BOUNDARIES	NUMBER IN CELL
4.395 to 4.495	2
4.495 to 4.595	2
4.595 to 4.695	8
4.695 to 4.795	15
4.795 to 4.895	14
4.895 to 4.995	13
4.995 to 5.095	16
5.095 to 5.195	15
5.195 to 5.295	11
5.295 to 5.395	3
5.395 to 5.495	1
Total Count	100

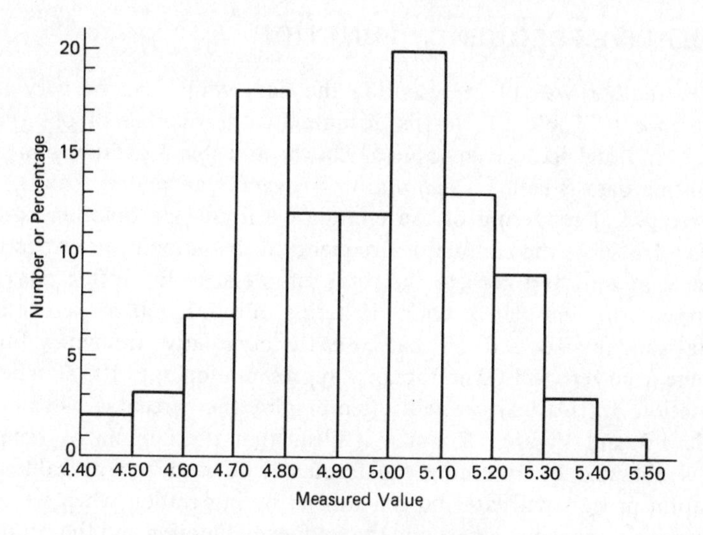

Figure 1.1. Histogram of Measurements.

Note that the vertical axis may represent the actual count in a cell or it may state the percentage of observations in the total sample occurring in that cell. Also, the range here was divided by the number 10 to generate a cell width, but any convenient number (usually between 8 and 20) could be used. Too small a number would not reveal the shape of the data and too large a number would result in many empty cells and a flat appearing distribution. Sometimes a few tries are required to arrive at a suitable choice.

In summary, the histogram provides us with a picture of the data from which we can intuitively see the center of the distribution, the spread, and the shape. The shape is important because we usually have an underlying idea or model as to how the entire population should look. The sample shape either confirms this expectation or gives us reason to question our assumptions. In particular, a shape that is symmetric about a center, with most of the observations in the central region, might reflect data from certain symmetric distributions, like the normal or Gaussian distribution. Alternatively, a nonsymmetric appearance would imply the existence of data points spaced farther from the center in one direction than in the other.

For the data presented in the fuse example, we note that the distribution appears reasonably symmetric. Hence, based on the histogram and the way the ends of the distribution taper off, the manufacturer believes that values much greater or much less than about 10% of the central target are not likely to occur. This variability he accepts as reasonable.

CUMULATIVE FREQUENCY FUNCTION

There is another way of representing the data which can be very useful. By reference to Table 1.2, let us accumulate the number of observations less than or equal to a given value as shown in Table 1.3. Such a means of representing data is called a *cumulative frequency function.*

The graphical rendering of the cumulative frequency function is shown as Figure 1.2. Note the cumulative frequency distribution is never decreasing and starts at zero and goes to the total sample size. It is often convenient to represent the cumulative count in terms of a fraction or percentage of the total sample size used. In that case, the cumulative frequency function will range from zero to 1.00 in fractional representation or to 100% in percentage notation. In this text, we will often employ the percentage form.

Table 1.3 and Figure 1.2 make it clear that the cumulative frequency curve is obtained by summing the frequency function count values. This summation process will later be generalized by integration when we discuss the population concepts underlying the frequency function and the cumulative frequency function in the next section.

THE CUMULATIVE DISTRIBUTION FUNCTION AND THE PROBABILITY DENSITY FUNCTION

The frequency distribution and the cumulative frequency distribution are calculated from sample measurements. Since the samples are drawn from a population, what can we state about this population? The typical procedure

Table 1.3. Cumulative Frequency Function.

UPPER CELL BOUNDARY (UCB)	NUMBER OF OBSERVATIONS LESS THAN OR EQUAL TO UCB
4.495	2
4.595	4
4.695	12
4.795	27
4.895	41
4.995	54
5.095	70
5.195	85
5.295	96
5.395	99
5.495	100

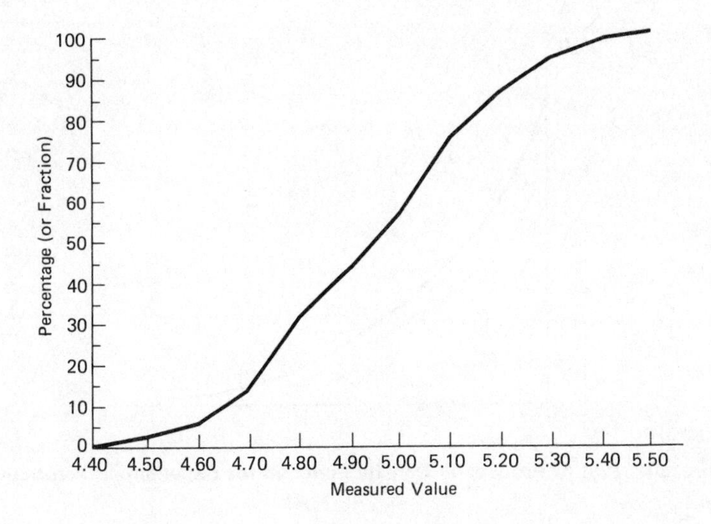

Figure 1.2. Plot of Cumulative Frequency Function.

is to assume a mathematical formula that provides a theoretical model for describing the way the population values are distributed. The sample histograms and the cumulative frequency functions are then estimates of these population models.

The model corresponding to the frequency distribution is the *probability density function* (PDF), denoted by $f(x)$ where x is any value of interest. The PDF may be interpreted in the following way: $f(x)\,dx$ is the fraction of the population values occurring in the interval dx. In reliability work, we often have time t as the variable of interest. So $f(t)\,dt$ is the fraction of failure times of the population occurring in the interval dt. A very simple example for $f(t)$ is called the exponential distribution given by the equation

$$f(t) = \lambda e^{-\lambda t}, \qquad 0 \leq t < \infty,$$

where λ is a constant. The plot of $f(t)$ is shown as Figure 1.3. The exponential distribution is a widely applied model in reliability studies and forms the basis of Chapter 3.

The cumulative frequency distribution similarly corresponds to a population model called the *cumulative distribution function* (CDF), denoted by $F(x)$. The CDF is related to the PDF via the following relationship

$$F(x) = \int_{-\infty}^{x} f(y)\,dy,$$

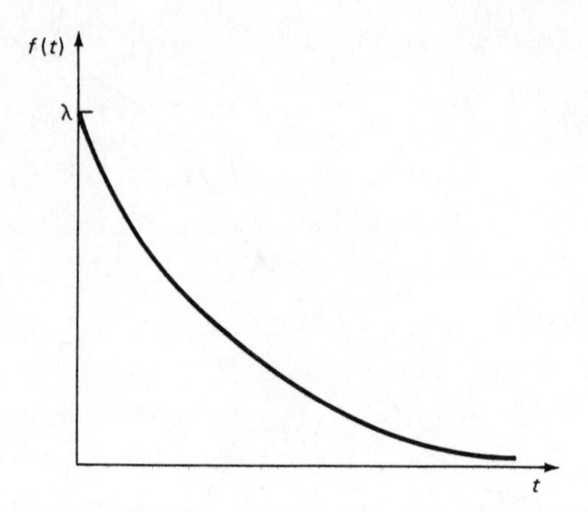

Figure 1.3. Plot of Probability Density Function for Exponential Distribution.

where y is the dummy variable of integration. $F(x)$ may be interpreted as the fraction of values in the population less than or equal to x. Alternatively, $F(x)$ gives the probability of a value less than or equal to x occurring in a single random draw from the population described by $F(x)$. Since in reliability work we usually deal with failure times, t, which are nonnegative, the CDF for population failure times is related to the PDF by

$$F(t) = \int_0^t f(y)\, dy, \qquad 0 \leq t < \infty.$$

For the exponential distribution,

$$F(t) = \int_0^t \lambda e^{-\lambda y}\, dy = -e^{-\lambda y}\Big]_0^t = 1 - e^{-\lambda t}.$$

The CDF for the exponential distribution is plotted as Figure 1.4.

When we calculated the cumulative frequency function in the fuse example, we worked with grouped data (data classified by cells). However, another estimate of the population CDF could have been generated by ordering the individual measurements from smallest to largest and then plotting the successive fractions

$$\frac{1}{n}, \frac{2}{n}, \frac{3}{n}, \ldots, \frac{n}{n}$$

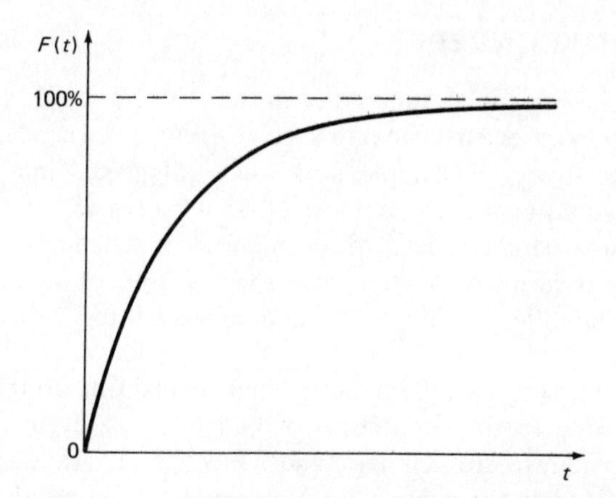

Figure 1.4. The CDF for the Exponential Distribution.

versus the ordered points. Such a representation is called the *empirical distribution function* (EDF) and is shown in Figure 1.5 for the data from the fuse example. The advantage of using the empirical distribution function instead of grouping the data is obviously that all data points are pictured; the disadvantage is that more computational effort is involved. However, computers are often available to perform the calculations.

Since $F(x)$ is a probability, all the rules and formulas for manipulating probabilities can be used when working with CDFs. Some of these basic rules will be described in the next section.

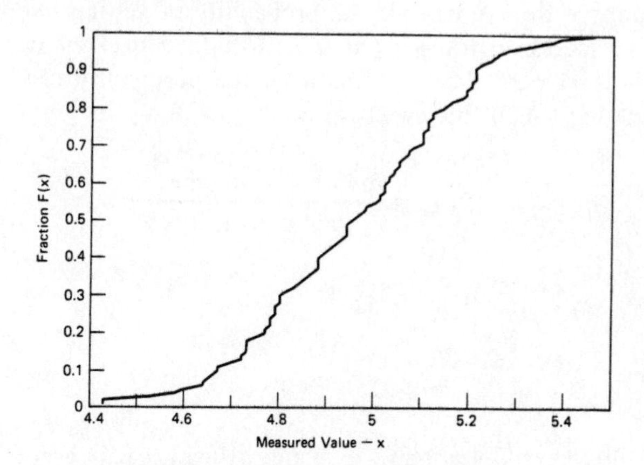

Figure 1.5. EDF for Fuse Data.

PROBABILITY CONCEPTS

The term probability can be thought of in the classical sense as the expected relative frequency of occurrence of a specific event in a very large collection of possible outcomes. For example, if we toss a balanced coin a large number of times, we would expect the number of occurrences of the event "heads" to comprise approximately half of the number of outcomes. Thus, we say the probability of heads on a single toss is 0.5 or 50% or 50–50. It is typical to express probabilities as either a fraction between 0 to 1 or as a percentage between 0 and 100.

There are two very useful relations often invoked in probability theory. These rules relate to the occurrence of two or more events. In electrical engineering terms, we are defining "and" and "or" relations. The first rule states that if $P(A)$ is the probability of event A occurring and $P(B)$ is the probability of event B occurring, then the probability of events A and B occurring simultaneously, denoted $P(AB)$, is

$$P(AB) = P(A)P(B|A)$$

or

$$P(AB) = P(B)P(A|B),$$

where $P(A|B)$ is the designator for the "conditional" probability of A given that event B has occurred.

Let us explain conditional probability further. We imply by the terminology that one event may be affected by the occurrence of another event. For example, suppose we ask what is the probability of getting two black cards in a row in successive draws from a well-shuffled deck of cards, without replacing the first card drawn. Obviously, the probability of the first card being a black card (call this event A) is

$$\begin{aligned} P(A) &= \frac{\text{favorable outcomes}}{\text{total outcomes}} \\ &= \frac{26}{52} \\ &= \frac{1}{2}. \end{aligned}$$

The probability of the second card being a black card (event B) changes depending on whether or not the first card drawn was a black card. If yes, then the probability of the second card being a black card is

$$P(B|A)=\frac{25}{51}.$$

So the probability of two successive black cards is

$$\begin{aligned}P(AB)&=P(A)P(B|A)\\&=\frac{1}{2}\frac{25}{51}\\&=\frac{25}{102}.\end{aligned}$$

If two events are independent, then the occurrence of one does not affect the probability of the other happening. In this case, for independent events A and B,

$$P(AB)=P(A)P(B).$$

In general, the probability of independent events occurring is just the product of the individual probabilities for each event. For example, in the card situation replacing the first card drawn and reshuffling the deck will make event B independent of event A. Thus, the probability of two successive black cards, with replacement and reshuffling between draws, is

$$\begin{aligned}P(AB)&=P(A)P(B)\\&=\frac{26}{52}\frac{26}{52}\\&=\frac{1}{4}.\end{aligned}$$

Similarly, the probability of simultaneously getting a 6 on one roll of a die and an ace in one draw from a deck of cards, apparently independent events, is

$$\begin{aligned}P(AB)&=\frac{1}{6}\frac{4}{52}\\&=\frac{1}{78}.\end{aligned}$$

The extension of these principles to three or more events is possible. For example, the rule for the joint probability of three events, A, B, and C, is

$$P(ABC) = P(A)P(B|A)P(C|AB).$$

For independent events, the formula becomes

$$P(ABC) = P(A)P(B)P(C).$$

The second important probability formula relates to the situation in which either of two events, A or B, may occur. The expression for this "union" is

$$P(A \cup B) = P(A) + P(B) - P(AB).$$

If the events are independent, then the relation becomes

$$P(A \cup B) = P(A) + P(B) - P(A)P(B).$$

The last term in the above expressions corrects for double counting of the same outcomes. For example, what is the probability of getting either an ace (event A) or a spade (event B) in one draw from a deck of cards? The events are independent, and so

$$\begin{aligned} P(A \cup B) &= P(A) + P(B) - P(A)P(B) \\ &= \frac{4}{52} + \frac{13}{52} - \frac{4}{52}\frac{13}{52} \\ &= \frac{16}{52} = \frac{4}{13}. \end{aligned}$$

Note the term $P(A)P(B)$ subtracts out the probability for a spade ace. This probability has already been counted twice, once in the $P(A)$ term and once in the $P(B)$ term.

When events A and B are mutually exclusive or disjoint, that is, both events cannot occur simultaneously, then $P(AB) = 0$, and

$$P(A \cup B) = P(A) + P(B).$$

Furthermore, if both events are also exhaustive in that at least one of them must occur when an experiment is run, then

$$P(A \cup B) = P(A) + P(B) = 1.$$

Thus, event A is the complement of event B. Event B can be viewed as the nonoccurrence of A and designated as event $\bar{A}$. Hence, the probability of occurrence of any event is equal to 1 minus the probability of occurrence of its complementary event. This complement rule has important applications in reliability work because a component may either fail (event A) or survive (event $\bar{A}$), resulting in

$$P(\text{failure}) = 1 - P(\text{survival}).$$

An extension to three or more events is also possible. For three events, A, B, and C, the formula is

$$P(A \cup B \cup C) = P(A) + P(B) + P(C) - P(AB) - P(BC) - P(AC) + P(ABC).$$

For independent events, the relation becomes

$$P(A \cup B \cup C) = P(A) + P(B) + P(C) - P(A)P(B) - P(B)P(C) - P(A)P(C) + P(A)P(B)P(C).$$

For mutually exclusive, exhaustive events, we have

$$P(A \cup B \cup C) = P(A) + P(B) + P(C) = 1.$$

RANDOM VARIABLES

In reliability studies, the outcome of an experiment may be numerical (e.g., time to failure of a component) or the result may be other than numerical (e.g., type of failure mode associated with a nonfunctional device). In either case, analysis is made possible by assigning a number to every point in the space of all possible outcomes, called the sample space. Examples of assigning numbers are: the time to failure is assigned the elapsed hours of operation; the failure mode may be assigned a category number 1, 2, and so on. Any rule for assigning a number creates a *random variable.*

> *A random variable is a function for assigning real numbers to points in the sample space.*

The practice is to denote the random variable by a capital letter (X, Y, Z, etc.) and the realization of the random variable (i.e., the real number or piece of sample data) by the lowercase letter (x, y, z, etc.). Since this definition appears a bit abstract, let us consider a simple example using a single die, with six faces, each face having one to six dots. The experiment consists of rolling the die and observing the upside face. The random variable is denoted X and assigns numbers matching the number of dots on the upside face.

Thus, $(X = x)$ is an *event* in the sample space and $X = 6$ refers to the realization where the face with six dots is upside. It is also common to refer to the probability of an event occurring using the notation $P(X = x)$. In this example, we usually assume all six possible outcomes are equally likely (fair die), and therefore, $P(X = x) = 1/6$ for $x = 1, 2, 3, 4, 5$, or 6.

SAMPLE ESTIMATES OF POPULATION PARAMETERS

We have discussed descriptive techniques such as histograms to represent observations. However, in order to complement the visual impression given by the frequency histogram, we often employ numerical descriptive measures called *parameters* for a population and *statistics* for a sample. These measures summarize the data in a population or sample and also permit quantitative statistical analysis. In this way, the concepts of central tendency, spread, shape, symmetry, and so on, take on quantifiable meanings.

For example, we stated that the frequency distribution was centered about a given value. This central tendency could be expressed in several ways. One simple method is just to cite the most frequently occurring value, called the *mode.* For grouped data, the mode is the midpoint of the interval with the highest frequency. For the fuse data in Table 1.1, the mode is 5.05.

Another procedure involves selecting the *median,* that is, the value that effectively divides the data in half. For individual readings the n data points are first ranked in order, from smallest to largest, and the median is chosen according to the following algorithm: the middle value if n is odd; the average of the two middle values if n is even. For grouped data, the median occurs in the interval for which the cumulative frequency distribution registers 50%, that is, a vertical line through the median divides the histogram into two equal areas. For grouped data with n points, to get the median one first determines the number of observations in the class containing the middle measurement $(n/2)$ and the number of observations in the class to get to that measurement. For example, for the fuse data $n = 100$, and the middle value is the fiftieth point. The fiftieth point occurs in the class marked 4.895 to 4.995 (width 0.1). There are 13 points in this class, and we must count to the ninth value to get to the middle value. Hence, the median is

$$4.895 + \frac{9}{13} \times 0.1 = 4.964.$$

(In reliability work, it is common terminology to refer to the median as the "T_{50}" value for time to 50% failures.)

The most common measure of central tendency, however, is called the arithmetic *mean* or average. The sample mean is simply the sum of the observations divided by the number of observations. Thus, the mean, denoted by $\bar{X}$, of n readings is given by the statistic

$$\bar{X} = \frac{X_1 + X_2 + X_3 + \cdots + X_n}{n},$$

$$= \frac{\sum_{i=1}^{n} X_i}{n}.$$

This expression is called a statistic because its value depends on the sample measurements. Thus, the sample mean will change with each sample drawn, which is another instance of the variability of the real world. The sample mean estimates the population mean. In contrast, the population mean depends on the entire set of measurements and thus is a fixed quantity which we call a parameter.

For a discrete (i.e., countable) population, the mean is just the summation over all discrete values where each value is weighted by the probability of its occurrence. For a continuous (i.e., measurable) population, the mean parameter is expressed in terms of the PDF model as

$$\mu = \int_{-\infty}^{\infty} x f(x)\, dx.$$

For reliability work involving time, the population mean is

$$\mu = \int_{0}^{\infty} t f(t)\, dt.$$

A common practice in statistics is to refer to the mean for both discrete and continuous random variables as the expected value of the random variable and use the notation $E(X) = \mu$ or $E(T) = \mu$. We shall occasionally use this terminology in this text.

Knowing the center of the distribution is not enough; we are also concerned about the spread of the data. The simplest concept for variability is the *range,* the difference between the highest and lowest readings. However, the range does not have very convenient statistical properties, and so another measure of dispersion is more frequently used. This numerical measure of variation is called the *variance.* The variance has certain statistical properties which make it very useful for analysis and theoretical work. The variance of a random variable X is defined as the expected value of $(X - \mu)^2$, that is, $V(x) = E[(X - \mu)^2]$. For continuous data, the population variance for common reliability analysis involving time is

$$V(t) = \sigma^2 = \int_0^\infty (t - \mu)^2 f(t)\, dt.$$

In engineering terms, we see that the variance is the expected value of the second moment about the mean.

The square root of the variance is called the *standard deviation.* The standard deviation is expressed in the same units as the observations. The sample variance is denoted by S^2 and the calculating formula is

$$S^2 = \frac{\sum_{i=1}^{n} (X_i - \overline{X})^2}{n - 1}.$$

The $n - 1$ term occurs because statistical theory shows that dividing by $n - 1$ gives a better (i.e., unbiased) estimate of the population variance (denoted by σ^2) than just dividing by n.

We have defined numerical measures of central tendency ($\overline{X}$, μ) and dispersion (S^2, σ^2). It is also valuable to have measures of symmetry about the center and a measure of how peaked the data are over the central region. These measures are called *skewness* and *kurtosis,* and are respectively defined as the expected values of the third and fourth moments about the mean, that is,

$$\text{skewness: } \mu_3 = E[(X - \mu)^3] \quad \text{and} \quad \text{kurtosis: } \mu_4 = E[(X - \mu)^4].$$

Symmetric distributions have skewness equal to zero. A unimodal (i.e., single-peak) distribution with an extended right "tail" will have positive skewness and will be referred to as skewed right; skewed left implies a negative skewness and a corresponding extended left tail. For example, the exponential distribution in Figure 1.3 is skewed right. Kurtosis, on the other hand, indicates the relative flatness of the distribution or how "heavy" the tails are. Both measures are usually expressed in relative (i.e., independent of the scale of measurement) terms by dividing μ_3 by σ^3 and μ_4 by σ^4. For further information, consult the text by Hahn and Shapiro (1967).

These various measures allow us to check the validity of assumed models. Ott (1977) shows applications to the normal distribution. Table 1.4 contains a listing of properties of distributions frequently used in reliability studies.

Table 1.4. Properties of Distributions Frequently Used in Reliability Studies.

PROPERTY	UNIFORM	NORMAL	WEIBULL	EXP	LOGNORMAL
Symmetric	Yes	Yes	No	No	No
Bell shaped	No	Yes	No	No	No
Skewed	No skew = 0	No skew = 0	Yes (right)	Yes (right) skew = 2	Yes (right)
Kurtois	1.8	3		9	
Log data Symmetric and bell shaped	No	No	No	No	Yes
Cumulative distribution	Straight line	"S" Shape		Exponential curve	

Two other useful distributions are:
Rayleigh: skew = 0.63, kurtosis = 3.26, a Weibull with shape = 2 and a linear failure rate.
Extreme value: Skew = −1.14 (skewed left), kurtosis = 5.4.

The important statistical concept involved in sample estimates of population parameters (e.g., mean, variance, etc.) is that the population parameters are *fixed* quantities, and we infer what they are from the sample data. For example, the fixed constant in the exponential model can be shown to be the mean of the distribution of failure times for an exponential population. The sample quantities, on the other hand, are random statistics which may change with each sample drawn from the population.

We also mention here a notation common in statistics and reliability work. The sample estimate of a population parameter is often designated by the symbol "^" over the population quantity. Thus, $\hat{\mu}$ is an estimate of the population mean μ; $\hat{\sigma}^2$ estimates σ^2, the population variance.

EXAMPLE 1.2 THE UNIFORM DISTRIBUTION

The uniform distribution is a continuous distribution with probability density function for the random variable T given by

$$f(t) = \frac{1}{\theta_2 - \theta_1}, \qquad \theta_1 \leq t \leq \theta_2,$$

and zero elsewhere, where θ_1 and θ_2 are the parameters specifying the range of T. The rectangular shape of this distribution is shown in Figure 1.6.

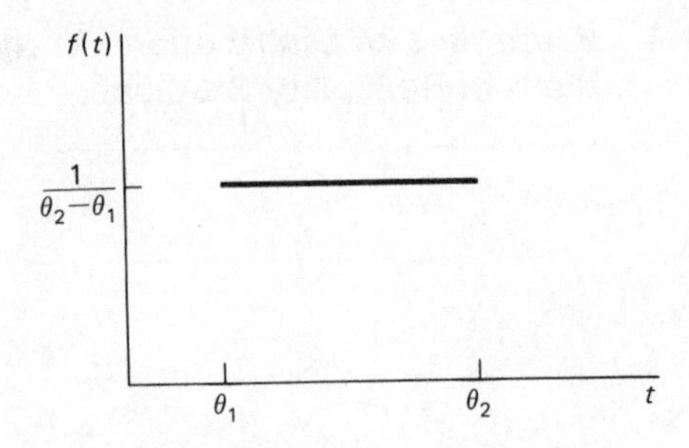

Figure 1.6. The Uniform PDF.

We note that $f(t)$ is constant between θ_1 and θ_2. The CDF of T, denoted by $F(t)$, for the uniform case is given by

$$F(t) = \frac{t - \theta_1}{\theta_2 - \theta_1}.$$

Thus, $F(t)$ is linear in t in the range $\theta_1 \leq t \leq \theta_2$, as shown in Figure 1.7. It is left to the reader to show that the uniform distribution has expected value $E(t) = (\theta_1 + \theta_2)/2$ and variance $V(t) = (\theta_2 - \theta_1)^2/12$.

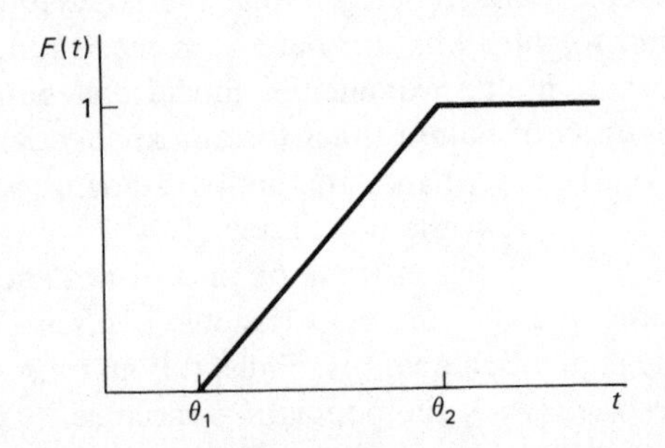

Figure 1.7. The CDF for the Uniform Distribution.

HOW TO USE DESCRIPTIVE STATISTICS

At this point it is important to emphasize some considerations for the analyst. No matter what summary tools or computer programs are available, the researcher should always "look" at the data, preferably in several ways. For example, many data sets can have the same mean and standard deviation and still be very different—and that difference may be of critical significance. See Figure 1.8 for an illustration of this effect.

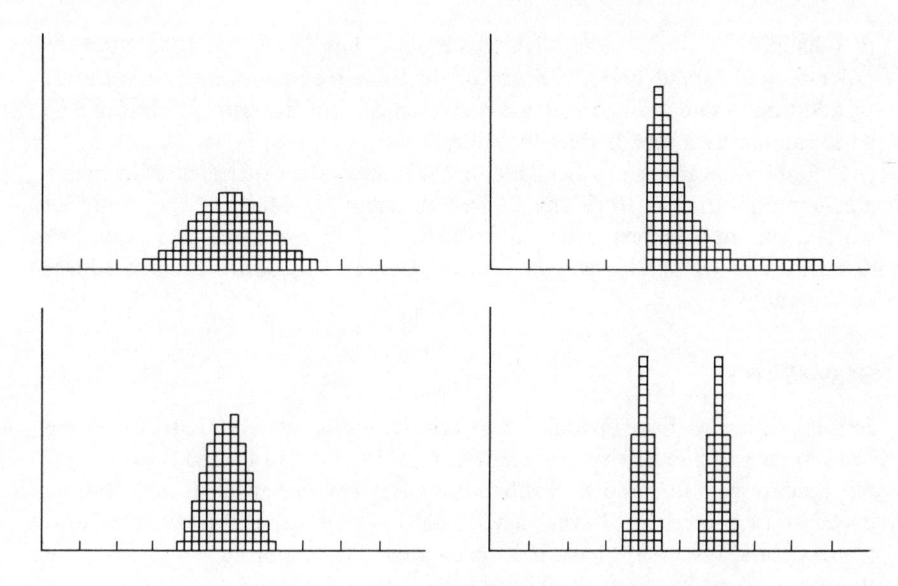

Distributions with the same mean and sigma

Figure 1.8. Mean and Sigma Do Not Tell Us Enough.

Generally the analyst will start out with an underlying model in his mind based on the type of data, where the observations came from, previous experience, familiarity with probability models, and so on. However, after obtaining the data it is necessary that the analyst go through a verification stage before he blindly plunges ahead with his model. This requirement is where the tools of descriptive statistics are very useful. Indeed, in many cases we utilize descriptive statistics to help us choose an appropriate model right at the start of our studies.

In this text, we will focus on several key distributions that are most applicable to reliability analysis. These are: the exponential, the Weibull, the normal, and the lognormal distributions. By learning what these distributions should look like, we can develop a yardstick by which to measure our data for appropriateness. Graphics (frequency histograms, cumulative frequency curves) and numbers (mean, median, variance, skewness, etc.) are the means by which the characteristics of distributions are understood. In later chapters, we shall introduce other valuable descriptive procedures such as probability plotting.

DATA SIMULATION

Many different PDFs (and CDFs) exist and reliability studies are often concerned with determining what model is most appropriate for the analysis.

In reliability work one may wish to simulate data from any distribution in order to understand better the use of different techniques and procedures. In addition, simulation can be a very valuable tool in terms of showing us what sample data results may look like.

Computer programs are available that will allow the experimenter to generate random variables from any desired distribution. Many of the examples worked out in this text will contain sets of data simulated from our key distributions. We shall use such simulated data to illustrate our procedures and methods.

SUMMARY

In this chapter we have introduced descriptive statistical techniques including histograms and cumulative frequency curves. We have discussed the concepts of populations and samples. Probability rules have been illustrated. Simple concepts of probability have been treated. Numerical measures of central tendency and dispersion have been presented. The importance of visualizing the observations has been emphasized. We have mentioned several important reliability distribution models. The next chapter will begin an in-depth presentation of the applications and uses of these concepts in the study of reliability.

2

Reliability Concepts

This chapter introduces the terms and concepts needed to describe and evaluate product reliability. These are the reliability function, the hazard and cumulative hazard function, the failure rate and average failure rate, the renewal rate, the mean time to failure, and the well-known "bathtub curve." In addition, we will look at the kinds of data a reliability analyst typically obtains from laboratory testing or a customer environment (uncensored, censored, and multicensored data).

THE RELIABILITY FUNCTION

The theoretical population models used to describe device lifetimes are known as "life distributions." For example, if we are interested in a particular type of transistor, then the population might be all the lifetimes obtainable from transistors of this type. Alternatively, we might want to restrict our population to just transistors from one particular manufacturer made during a set period of time. In either case, the CDF for the population is called a life distribution. If we denote this CDF by $F(t)$, then $F(t)$ has two useful interpretations.

1. $F(t)$ is the probability a random unit drawn from the population fails by t hr.
2. $F(t)$ is the fraction of all units in the population which fail by t hr.

Pictorially, $F(t)$ is the area under the probability density function $f(t)$ to the left of t. This area is shown in Figure 2.1. The total area under $f(t)$ is unity (i.e., the probability of failure approaches 1 as t approaches infinity).

Since $F(t)$ is a probability, the shaded region has an area equal to the probability of a new unit failing by t hr of operation. This equivalence of area to probability generalizes so that the area under $f(t)$ between two vertical lines drawn at time t_1 and a later time t_2 corresponds to the probability of

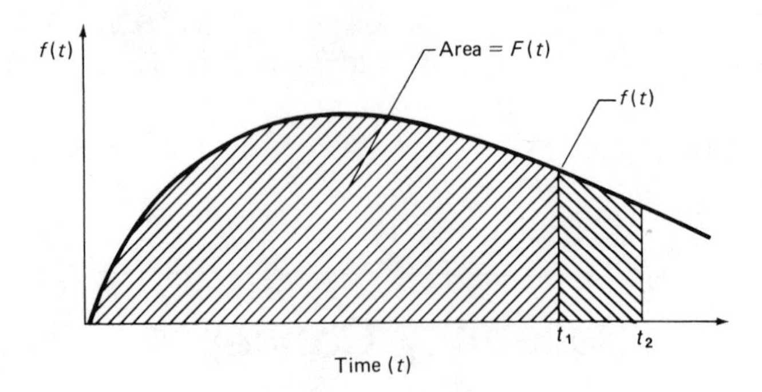

Figure 2.1. CDF *F(t)*.

a new unit surviving to time t_1 and then failing in the interval between t_1 and t_2. This area can be obtained by taking all the area to the left of t_2 and subtracting the area to the left of t_1. But this is just $F(t_2)-F(t_1)$.

> $F(t_2)-F(t_1)$ is the probability a new unit survives to time t_1 but fails before time t_2. It is also the fraction of the entire population which fails in that interval.

Since it is often useful to focus attention on the unfailed units, or survivors, we define the reliability function (survival function) by

$$R(t) = 1 - F(t)$$

(We could also call $F(t)$ the "unreliability" function.)

The reliability function may be thought of in either one of two ways:

1. As the probability a random unit drawn from the population will still be operating after t hours.
2. As the fraction of all units in the population that will survive at least t hours.

If n identical units are operating and $F(t)$ describes the population they come from, then $nF(t)$ is the expected (or average) number of failures up to time t and $nR(t)$ is the number expected to be still operating.

EXAMPLE 2.1 LIFE DISTRIBUTION CALCULATIONS

Suppose that a population of components is described by the life distribution, $F(t) = 1 - (1 + .001t)^{-1}$. What is the probability a new unit will fail by

1000 hr? By 4000 hr? Between 1000 and 4000 hr? What proportion of these components will last more than 9000 hr? If we use 150 of them, how many do we expect to fail in the first 1000 hr? In the next 3000 hr?

SOLUTION

By substitution, $F(1000) = 1 - (1 + 1)^{-1} = .5$, and $F(4000) = 1 - (1 + 4)^{-1} = .8$. These are the probabilities of failing in the first 1000 and first 4000 hr, respectively. The probability of failing between 1000 and 4000 hr is $F(4000) - F(1000)$, or .3. The proportion surviving past 9000 hr, $R(9000)$, is $(1 + 9)^{-1}$, or .1. Finally, the expected failures in the first 1000 hr are $150 \times .5$, or 75. In the next 3000 hr, an additional $150 \times .3$, or 45, are expected to fail.

SOME IMPORTANT PROBABILITIES

Since $F(t)$ and $R(t)$ are probabilities, a few simple but powerful formulas can be derived without difficulty by using the basic rules for calculating probabilities of events as presented in Chap. 1. The two rules needed in this section are as follows:

1. Multiplication Rule: The probability that several independent events will all occur is the product of the individual event probabilities.
2. Complement Rule: The probability that an event will not occur is 1 minus the probability of the event.

In our application, the independent events are the failure or survival of each of n randomly chosen, independently operating units. If we want the probability of all of them still being in operation after t hours, we must apply the Multiplication Rule and multiply $R(t)$ by $R(t)$ n times. In other words,

> The probability that n independent identical units, each with reliability $R(t)$, will all survive after t hours is $[R(t)]^n$.

If we want the probability that at least one of the n units will fail, we must apply the Complement Rule and obtain:

> The probability that at least one of n independent identical units fails by time t is given by
>
> $$1 - [R(t)]^n = 1 - [1 - F(t)]^n.$$

The power of these formulas is readily apparent if we consider a simple system composed of n identical components all operating independently, in terms of working or failing, of each other. If the life distribution for each of these components is $F(t)$, then the probability the system does not have a failure by time t is $[R(t)]^n$. If the system fails when the first of its components fails, and we denote the life distribution function for a population of these systems by $F_s(t)$, then the complement rule gives us

$$F_s(t) = 1 - [R(t)]^n.$$

This equation shows how, in this simple case, system reliability is built up using a bottoms up approach starting with the individual component reliabilities. This is a key concept, which will be discussed in detail in Chapter 8.

We turn now from the probabilities of failure to the various ways of defining rates at which failures occur.

THE HAZARD FUNCTION OR FAILURE RATE

Consider a population of 1000 units that start operating at time zero. Over time, units fail one by one. Say that at 5000 hours the fourth unit has already failed and another unit fails in the next hour. How would we define a "rate of failure" for the units operating in the hour between 5000 and 5001? Since 996 units were operating at the start of that hour and one failed, a natural estimate of the failure rate (for units at 5000 hr of age) would be $1/996$ per hour.

If we look closely at that calculation, we see that we have calculated a conditional rate of failure, or rate of failure for the survivors at time t.

We can make this definition more precise by using the concept of conditional probability discussed in Chapter 1. There, we used the notation $P(A)$ to denote the probability of event A occurring and the notation $P(B|A)$ to denote the conditional probability that event B will occur given that A is known to have occurred. $P(B|A)$ was defined as follows:

$$P(B|A) = P(B \text{ and } A \text{ both occur}) \div P(A)$$

Using this formula, we can calculate the probability of failing after surviving up to time t in a small interval of time Δt as follows:

$$P(\text{fail in next } \Delta t | \text{survive to } t) = \frac{F(t + \Delta t) - F(t)}{R(t)}$$

We divide this by Δt to convert it to a rate and obtain

$$\frac{F(t+\Delta t)-F(t)}{R(t)\,\Delta t}$$

If we now let Δt approach zero, we obtain the derivative of $F(t)$, denoted by $F'(t)$, divided by $R(t)$. Since $F'(t)=f(t)$, we have derived the expression for the instantaneous failure rate, or hazard rate $h(t)$:

$$h(t)=\frac{f(t)}{R(t)}$$

For the remainder of the text, the terms *failure rate, instantaneous failure rate,* and *hazard rate* will all be equivalent and have the above definition.

The rate we have just defined has units of failures per unit time. It is the failure rate of the survivors to time t in the very next instant following t. It is not a probability and can have values greater than 1 (although it is always nonnegative). In general, it is a function of t and not a single number or constant.

The reader should be cautioned that not all authors use the same definition when talking about failure rates. Some authors define the failure rate to be $f(t)$, which is the rate of failure of the original time zero population at time t.

THE CUMULATIVE HAZARD FUNCTION

Just as the probability density function $f(t)$ can be integrated to obtain the cumulative distribution function $F(t)$, we can integrate the hazard function $h(t)$ to obtain the cumulative hazard function $H(t)$.

$$H(t)=\int_0^t h(t)\,dt.$$

This integral can be expressed in closed form as

$$H(t)=-\ln R(t)$$

where the notation ln denotes natural logarithms or logarithms to the base e. [The reader can easily verify this equation by taking derivatives of both sides and obtaining $h(t)=f(t)/R(t)$.]

By taking antilogarithms in the above equation for $H(t)$, a well-known and useful identity relating failure rates and CDFs is obtained:

$$F(t) = 1 - e^{-H(t)} = 1 - e^{-\int_0^t h(t)\,dt}.$$

This shows that given $H(t)$, we can calculate $F(t)$, and vice-versa. So, in a sense, all the quantities we have defined give the same amount of information: with any one out of $F(t)$, $f(t)$, $h(t)$, or $H(t)$, we can calculate all of the others.

$H(t)$ will be particularly useful later when we discuss graphical plotting methods for estimating life distribution parameters from failure data.

THE AVERAGE FAILURE RATE

Since the failure rate $h(t)$ varies over time, it is useful to define a single average number that typifies failure rate behavior over an interval. This number might be used in an engineering specification for a component, or it might be an input to service cost and stock replacement calculations.

A natural way to define an average failure rate (AFR) between time t_1 and t_2 is to integrate the (instantaneous) failure rate over the interval and divide by $t_2 - t_1$.

$$\begin{aligned} \text{AFR}(t_1,t_2) &= \frac{\int_{t_1}^{t_2} h(t)\,dt}{t_2 - t_1} \\ &= \frac{H(t_2) - H(t_1)}{t_2 - t_1} \\ &= \frac{\ln R(t_1) - \ln R(t_2)}{t_2 - t_1}. \end{aligned}$$

If the time interval is from 0 to T, the AFR simplifies to

$$\text{AFR}(T) = \frac{H(T)}{T} = \frac{\ln R(T)}{T}$$

and this quantity is approximately equal to $F(T)/T$ for small $F(T)$.

This last AFR finds frequent use as a single-number specification for the overall failure rate of a component that will operate for T hr of useful life. For example, if the desired lifetime is 40,000 hr, then AFR(40,000) is the single average lifetime failure rate.

It should be noted that the AFR is not generally defined or used in most of the literature on reliability, despite its usefulness.

UNITS

Failure rates for components are often so small that units of failures per hour would not be appropriate. Instead, the scale most often used for failure rates is percent per thousand hours (%/K). One percent per thousand hours would mean an expected rate of 1 fail for each 100 units operating 1000 hr. Another scale rapidly becoming popular for highly reliable components is parts per million per thousand hours (PPM/K). One part per million per thousand hours means 1 fail is expected out of 1 million components operating for 1000 hr. Another name for PPM/K is FIT for fails in time. This name will be used for the rest of this text.

The factors to convert $h(t)$ and the AFR to %/K or FIT are given below:

$$\begin{aligned}
\text{failure rate in \%/K} &= 10^5 \times h(t) \\
\text{AFR in \%/K} &= 10^5 \times \text{AFR}(T_1, T_2) \\
\text{failure rate in PPM/K} &= 10^9 \times h(t) \\
\text{AFR in PPM/K} &= 10^9 \times \text{AFR}(T_1, T_2).
\end{aligned}$$

EXAMPLE 2.2 FAILURE RATE CALCULATIONS

For the life distribution in Example 2.1, derive $h(t)$ and calculate the failure rate at 10, 100, 1000, and 10,000 hr. Give the last failure rate in both PPM/K and FIT. What is AFR(1000)? What is the average failure rate between 1000 and 10,000 hr? If five components, each having this life distribution, are starting operation, what is the probability they have no failures in the first 1000 hr?

SOLUTION

$F(t) = 1 - (1 + .001t)^{-1}$ and, by taking the derivative, $f(t) = .001(1 + .001t)^{-2}$. By definition

$$h(t) = \frac{f(t)}{1 - F(t)} = .001(1 + .001t)^{-1}$$

By substitution, $h(10) = .001 \div 1.01 = .00099$. Similarly, $h(100) = .001/1.1 = .00091$ and $h(1000) = .0005$ and $h(10{,}000) = .000091$. $\text{AFR}(1000) = -\ln R(1000)/1000 = -\ln(1 + 1)^{-1}/1000 = .0007$. $\text{AFR}(1000{,}10000) = [\ln 2^{-1} - \ln 11^{-1}]/9000 = .00019$.

The probability that five components operate for 1000 hr without any failures is $[R(1000)]^5$ by applying the Multiplication Rule. This yields $[(1 + 1)^{-1}]^5 = 2^{-5} = 1/32$.

BATHTUB CURVE FOR FAILURE RATES

Any nonnegative function $h(t)$ whose integral $H(t)$ approaches infinity as time approaches infinity can be a failure rate function. However, in practice most components, or groups of components operating together as subassemblies or systems, tend to have failure rate curves with a similar kind of appearance.

This typical shape for failure rate curves, known as the bathtub curve, is shown in Figure 2.2. The first part of the curve, known as the early failure period, has a decreasing failure rate. During this period the weak parts that were marginally functional are weeded out. In Chapter 8 we will see that it also makes sense to include the discovery of defects that escape to the user in the front end of this curve.

The long fairly flat portion of the failure rate curve is called the stable failure period (also known as the intrinsic failure period). Here failures seem to occur in a random fashion at a uniform or constant rate. Most of the useful life of a component should take place in this region of the curve.

The final part of the curve, where the failure rate is increasing, is known as the wearout failure period. Here degradation failures occur at an ever increasing pace. One of the main purposes of reliability testing is to assure that the onset of the wearout period occurs far enough out in time as to not be a concern during the useful life of the product. (The other key purpose of reliability testing is to establish the value of the failure rate during the long stable period.)

It is interesting to note that while every experienced reliability analyst has come across many examples of real failure data exhibiting the shape shown in Figure 2.2, none of the prominent life distributions discussed in the literature or in this book, have that shape for $h(t)$. These distributions, as we shall see in later chapters, fit only one or at most two regions of the

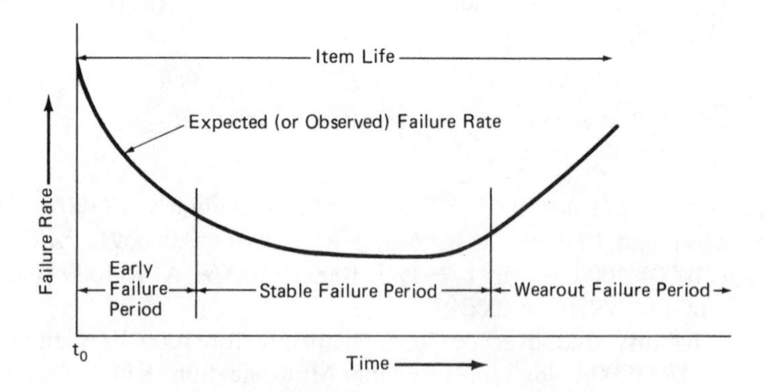

Figure 2.2. Bathtub Curve for Failure Rates.

curve reasonably well. In Chapter 8, algorithms for obtaining the entire curve will be developed.

RENEWAL RATES

Another kind of failure rate that takes into account the fact that failed parts are replaced with parts that may, in turn, later fail can be defined. If the replacements are always new parts from the same population as the original parts, we call the process a renewal process. The mean or expected number of failures per unit interval at time t is known as the renewal rate $r(t)$.

Which failure rate is better to use—the hazard rate we are calling the failure rate or the renewal rate? The hazard rate (or the AFR) corresponds to a process where, instead of replacing parts with brand new parts when they fail, a part similar to the failed part in terms of remaining life is used.

Typically, however, the replacement part is new but comes from a different time or vintage of manufacture than the original part. This means it may have a different failure rate curve and neither the renewal rate concept nor the failure rate we have defined applies exactly.

Fortunately, when failure rates are nearly constant, as on the long flat portion of the bathtub curve, it does not matter which rate is used; both the renewal rate, the failure rate, and the AFR have the same value when the failure rate is constant. For this reason, and because the renewal rate is very difficult to work with mathematically, we will use failure rates and AFRs predominantly throughout this text. (A method for estimating $F(t)$ from renewal data will be given in Chapter 8, however.)

EXAMPLE 2.3 EXAMPLES OF VARIOUS RATES

In Figure 2.3 10 similar components operating in a system are represented by horizontal lines. An X on the line shows that a failure occurred at the corresponding time. After each failure, the failing component is replaced by a new component and the system continues to operate. The first failure for each component is shown by ⓧ.

How would we estimate the CDF and reliability function at 500 and at 550 hr? What would an estimate of the failure rate or average failure rate be for the interval 500 to 550 hr? How does this compare to an estimate of the renewal rate based on the same interval? What would an estimate of $f(t)$ over this interval be?

SOLUTION

The various estimates derived from Figure 2.3 are

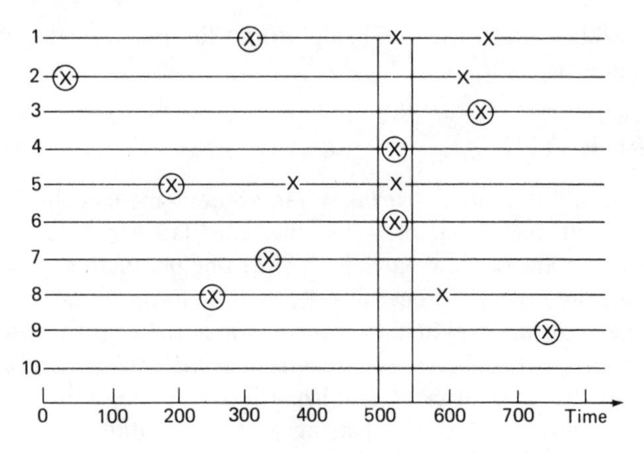

Figure 2.3. Component Failure Data Example.

$$\hat{F}(500) = 5/10$$
$$\hat{R}(500) = 5/10$$
$$\hat{F}(550) = 7/10$$
$$\hat{R}(550) = 3/10$$

$$\hat{h}(500 \text{ to } 550) = \frac{2/5}{50 \text{ hr}} = \text{AFR}(500{,}550)$$

$$\hat{r}(500 \text{ to } 550) = \frac{4/10}{50 \text{ hr}}$$

$$\hat{f}(500 \text{ to } 550) = \frac{2/10}{50 \text{ hr}}$$

This example clearly illustrates the differences between F, f, and the failure rate and renewal rate. To estimate F at a given time, the denominator is the starting number of components and the numerator is all the first time only failures. To estimate f over an interval, the same denominator is used, but only the first time failures in that interval are in the numerator. This estimate approaches the true f in value if the population is large and the interval is small.

The failure rate or $h(t)$ estimate uses a different denominator. Instead of the time zero original population, only the survivors at the start of the interval are used in the calculation. Since five of the original units had failed by 500 hr, the denominator is $10 - 5 = 5$. The numerator is the number of first time failures in the interval or 2. This is actually an estimate of the average failure rate over the interval; as with the estimate of $f(t)$, the larger the original population and the smaller the interval, the better an estimate of $h(t)$ this becomes.

The renewal rate estimate uses the total number of operating units in the denominator. Under the renewal concept, this will also equal the starting population. The numerator now includes all the failures in the interval, whether from the starting population (circled), or failures of replacement components (uncircled). This number is 4 in the example.

Note that all the rate estimates also have a division by 50 hr to convert to failures per unit hours. This could then be multiplied by the appropriate constant, as explained in the section on units, if %/K or FIT were desired.

TYPES OF DATA

The statistical analysis of reliability data is often more complicated and difficult than the analysis of other experimental data. This is because of the diversity of forms this data takes. Most other experiments yield straightforward sample data: a random sample of size n gives n numbers that can be used in a simple way to make inferences about the underlying population. This kind of data is seldom available in reliability evaluation.

The kinds of data generally encountered fall into the categories of exact failure times with censoring (Type I or Type II), readout (or interval or grouped) data, and the most general form of all—multicensored data.

Exact Times—Censored Type I

Suppose n units are put on test for a fixed planned duration of T hr. Say that r of them fail, and the exact failure times $t_1 \leq t_2 \leq t_3 \leq \cdots \leq t_r \leq T$ are recorded. At the end of the test there are $n - r$ survivors (unfailed units). All that is known about them is that the times of failure they will eventually record are beyond T.

This kind of testing is called censored Type I. It has the advantage of ensuring schedules are met, since the duration T is set in advance. This is valuable in situations where there may be serious impacts if schedules are not rigidly kept.

There is one problem with this kind of testing. The number of failures is not known in advance. This number, r, is a random quantity. Since, as we shall see, the precision of failure rate estimates depends on the number of failures r, and not on the number of units n on test, a bad choice of sample size or test conditions may result in insufficient information obtained from the test. The test time may fit within the allotted schedule, but the test results may be inadequate.

Exact Times—Censored Type II

Again, we place n units on test and record the exact times when failures occur. However, instead of ending the test at a predetermined time, we wait until exactly r failures occur and then stop. Since we specify r in advance, we know exactly how much data the test will yield.

This procedure has obvious advantages in terms of guaranteeing adequate data. However, the length of test time is random and open ended. Based on practical scheduling considerations, Type II testing is usually ruled out in favor of Type I.

Readout Time Data

Both above testing schemes require instruments that can record exact times of failure. Thus when testing electronic components, continuous in situ monitoring is needed. This kind of test setup may be impractical in many cases.

The following is a practical, commonly used testing procedure: n components are put on test at time zero. T_1 hr later, a "readout" takes place where all the components are examined and failures are removed. Say r_1 failures are found. Then $n - r_1$ components go back on test. At time T_2, after $T_2 - T_1$ more hours of test, another readout takes place. This time r_2 failures are found and $n - r_1 - r_2$ units go back on test. The process is continued with readouts at time T_3 and T_4, and so on. The last readout, at time $T_k = T$, takes place at the end of the test. Figure 2.4 illustrates this kind of data.

This type of data is called readout or interval or grouped data. The readout times are predetermined, as is the end of test time. How many failures will occur in an interval is not known until the readout takes place. The exact times of fail are never known.

Readout data experiments have the same problem censored type data has; the experiment may end before a sufficient number of failures take place. In addition, precision is lost by not having exact times of failure recorded. Even if there are many failures, the data may be inadequate if these failures are spread out over too few intervals.

Despite the above drawbacks, and the difficulties analyzing readout data, it is probably the most common type of reliability data. In situ monitoring usually requires expensive test equipment that is often not available nor cost justified. The reliability analyst must learn to make the best use possible of readout data experiments; this goal involves careful planning of the experiment (i.e., sample sizes and times of readout) as well as use of good analysis methods.

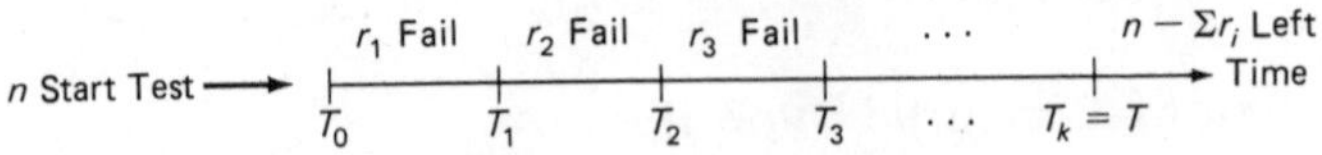

r_1 Fail r_2 Fail r_3 Fail . . . $n - \Sigma r_i$ Left

n Start Test → Time

T_0 T_1 T_2 T_3 . . . $T_k = T$

Figure 2.4. Readout Data.

Multicensored Data

In the most general case, every unit on test may have an interval specified for it during which it is known to have failed, or a censoring time it is known to have survived past. These intervals and censoring times might be different for each unit. An exact time of failure would be a "degenerate" or very small interval. This kind of data is called multicensored. In the laboratory, it is rare that every unit has a different readout time or censoring time, but less complicated examples of multicensored data are common. In the field, all types of multicensored data frequently occur.

EXAMPLE 2.4 MULTICENSORED DATA

Capacitors are to be tested on fixtures mounted in ovens. While the test is on, the parts will be subjected to a fixed high voltage and high temperature. At the end of each day, they will be removed from the ovens and tested to determine which ones have failed. Then the unfailed units will be put back on test. The plan is to continue like this for 1000 hr.

This test plan would normally yield standard readout data, but several unexpected things might happen to change this. Two possibilities will illustrate this point.

Assume the test starts with 200 capacitors in four ovens, each containing 50 of them. Halfway through the test one of the ovens malfunctions, causing all further data on its 50 parts to be invalid. This results in multicensored data with the capacitors in the bad oven "taken off test" at the time of malfunction. The other ovens and parts continue for the full test time.

Instead of a malfunction, the same situation might occur if priorities change and the test engineer must give up one of his ovens before his capacitor experiment is completed.

These examples show that a mild form of multicensoring may occur even when straightforward readout data was expected. The next example shows a case where heavily multicensored data arises as a natural consequence of the data collection scheme.

EXAMPLE 2.5 MULTICENSORED DATA

Data obtained from components operating in customer environments often have a characteristic form of multicensoring. The data, in this case, consist of failure information on several groups of components, each group on test for a possibly different interval of time and readout once at the end of that interval. A concrete example will make this clear.

Assume a machine uses 100 components of a type that are of interest. We are able to examine the field history of 10 machines. Each of these machines has operated a different length of time, based on the date of customer installation. For each machine, only the total number of component failures, up to the date of the investigation, is available. All failures are assumed to come from the original 100 components and either no repairs were necessary, despite failures, or repair time is considered negligible. (Error correction circuitry in computer memories is an example where a machine continues to operate despite component failures.)

Each machine provides data on a group of 100 components that were "on test" for just one readout interval. That interval is the amount of time the machine has operated and varies from machine to machine. There are also 10 different censoring times: one for each machine's group of unfailed units.

The kind of multicensored data described in this example does not lend itself to graphical methods and can be very difficult to analyze unless the exponential distribution (described in Chapter 3) applies.

FAILURE MODE SEPARATION

Multicensored data can also come about when test components fail for more than one reason. For example, a corrosion failure mode and a metal migration failure mode might both take place when testing a semiconductor chip. Each mode of failure might follow a completely different life distribution and it is desirable to analyze the data for each mode separately.

If the failure modes act independently of one another, we can analyze the corrosion data alone by treating the times when units failed due to migration as censoring times. After all, if migration failure had not occurred, these units would have eventually failed due to corrosion. Instead they were "taken off test" at the time of their migration failure.

The analysis of migration failures is done in a similar fashion, by treating the times of corrosion failures as censoring times. This approach can be extended to more than two independent modes of failure. This topic is treated further in the section on data analysis in Chapter 8.

SUMMARY

This chapter defined the reliability function $R(t) = 1 - F(t)$ and the (instantaneous) failure rate $h(t) = f(t)/R(t)$. This failure rate applies at the instant of time t for the survivors of the starting population still in operation at that time. An average failure rate over the interval (t_1,t_2) can also be defined by integrating the failure rate over this interval and dividing by $(t_2 - t_1)$. This AFR turns out to equal $[\ln R(t_1) - \ln R(t_2)]/(t_2 - t_1)$.

A useful identity that shows how the CDF can be reconstructed from the failure rate is

$$F(t) = 1 - e^{-\int_0^t h(t)dt}.$$

A plot of failure rate versus time for most components or systems of components yields a curve with the so-called "bathtub" shape. The front decreasing portion shows early life fallout. Then a long fairly flat region occurs. Finally, at some point in time wearout failures due to degradation mechanisms start to dominate and the failure rate rises.

One common aspect of reliability data causes analysis difficulties. This feature is the censoring that takes place because, typically, not all units on test fail before the test ends.

Reliability data may consist of exact times of failure up to the end of a fixed length test. This type of data is called Type I censored (also referred to as time censored). Type II censoring (also called failure censoring) refers to a test that lasts until a prespecified number of failures occur. While Type II censoring may lead to better data, it is less popular because of the open-ended nature of the test duration.

Perhaps the most typical kind of data consists of numbers of failures known only to have occurred between test readouts. This is called readout or interval or grouped data. If all the units do not have the same readout intervals or end of test times, the data are called multicensored and are the most difficult of all to analyze.

Multicensored data may also come about when some units on test are damaged or removed prematurely or fail due to more than one failure mode. Field (or customer operation) data can also come in a multicensored form.

3
The Exponential Distribution

The exponential distribution is one of the most common and useful life distributions. In this chapter, we will discuss the properties and areas of application of the exponential and then look at how to estimate exponential failure rates from data. Tables of factors to calculate upper and lower confidence bounds are included. Use of these tables to solve many important problems of experimental planning will be illustrated with detailed examples.

EXPONENTIAL DISTRIBUTION BASICS

The PDF for the exponential,

$$f(t) = \lambda e^{-\lambda t}$$

and the CDF,

$$F(t) = 1 - e^{-\lambda t}$$

were introduced in Chap. 1. Figures 1.3 and 1.4 gave plots of these functions.

In both of these equations, λ is the single unknown parameter that defines the exponential distribution. If λ is known, values of $F(t)$ can easily be calculated for any *t*. Only an inexpensive calculator or a table of natural logarithms is needed.

Since $R(t) = 1 - F(t) = e^{-\lambda t}$, the failure rate function [using the definition of $h(t)$ given in Chapter 2] for the exponential distribution is

$$h(t) = \frac{f(t)}{R(t)} = \frac{\lambda e^{-\lambda t}}{e^{-\lambda t}} = \lambda.$$

This result shows that the exponential failure rate function reduces to the value λ for all times. This is a characteristic property of the exponential

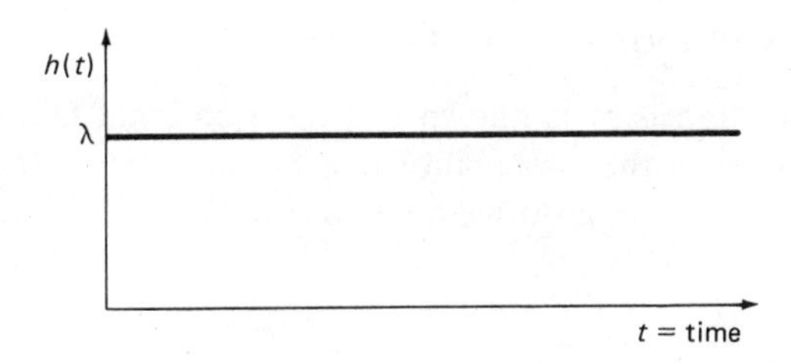

Figure 3.1. The Exponential Distribution Failure Rate $h(t)$.

distribution; the only distribution with a constant failure rate function is the exponential (see Example 3.2). Figure 3.1 shows how this failure rate looks when plotted against time.

The units for λ are failures per unit time—consistent with whatever units time is measured in. Thus, if time is in hours, λ is in failures per hour. If time is in thousand hour (abbreviated as K) units, then λ is in failures per K. Often, however, failure rates are expressed in percent per thousand hour units; this rate has to be converted to failures per unit time before making any calculations using the exponential formula. The same is true if failure rates are measured in FITs, which are failures per 10^9 device hours, as explained in the section on units in Chapter 2. Table 3.1 shows some examples of how to convert between these systems of units.

The average failure rate, or AFR, between time t_1 and time t_2 was defined in Chapter 2 to be $[\ln R(t_1) - \ln R(t_2)]/(t_2 - t_1)$. For the exponential, this expression turns out to be $\lambda(t_2 - t_1)/(t_2 - t_1)$ which reduces again to the constant λ.

Table 3.1. Equivalent Failure Rates in Different Units.

FAILURES PER HOUR	%/K	FIT
.00001	1.0	10,000
.000001	.1	1,000
.0000001	.01	100
.00000001	.001	10
.000000001	.0001	1

Failures per hour $\times 10^5$ = %/K
Failures per hour $\times 10^9$ = FIT
%/K $\times 10^4$ = FIT

EXAMPLE 3.1 EXPONENTIAL PROBABILITIES

A certain type of transistor is known to have a constant failure rate with $\lambda = .04\ \%/K$. What is the probability one of these transistors fails before 15,000 hr of use? How long do we have to wait to expect 1% failures?

SOLUTION

First we convert λ to failures per hour units by multiplying by 10^{-5}. This gives $\lambda = .0000004$. The probability of failure by 15,000 hr is $F(15{,}000)$ or $1 - e^{-.0000004 \times 15000} = .006 = .6\%$.

We find the time corresponding to any percentile by inverting the formula for the CDF [solving for t in terms of the proportion $F(t)$]. This inversion gives

$$t = [-\ln(1 - F(t)]/\lambda$$

and substituting .01 for $F(t)$ and .0000004 for λ gives $t = 25{,}126$ hr.

EXAMPLE 3.2 CONSTANT FAILURE RATE

Show that a constant failure rate implies an exponential distribution model.

SOLUTION

The basic identity relating failure rates to CDFs was derived in Chapter 2. The general formula is

$$F(t) = 1 - e^{-\int_0^t h(t)dt}.$$

If $h(t) = \lambda$, the integral becomes $\int_0^t \lambda\, dt = \lambda t$. Then we have $F(t) = 1 - e^{-\lambda t}$, or the exponential CDF.

The theoretical model for the shape we expect exponential data to resemble, when plotted in histogram form, is that of the PDF $f(t) = \lambda e^{-\lambda t}$. This was shown originally as Figure 1.3. We repeat the plot in Figure 3.2, for easy reference.

Sample data that are suspected of coming from an exponential distribution can be plotted in histogram form, as described in Chapter 1, and compared in shape to this ideal form. An alternative way of "looking at data" to see if it appears exponential will be discussed in Chapter 6 (graphical plotting).

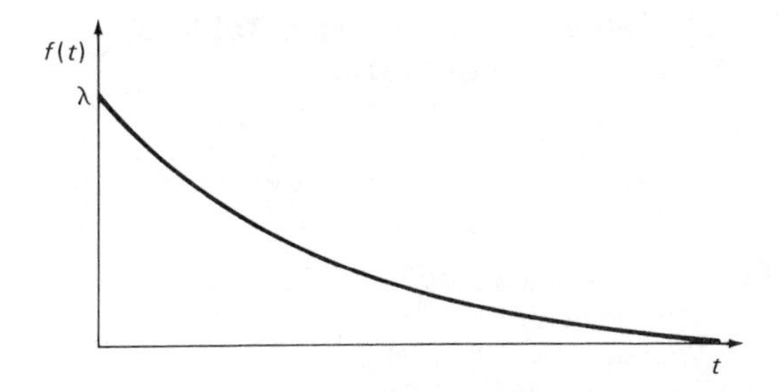

Figure 3.2. The Exponential PDF.

EXAMPLE 3.3 EXPONENTIAL DATA

The Lifetime Light Bulb company makes an incandescent filament that they believe does not wear out during an extended period of normal use. They want to guarantee it for 10 years of operation. The Quality Department is given three months to determine what such a guarantee is likely to cost.

Fortunately, the engineer who has to come up with a test plan has a verified way of stressing light bulbs (using higher than normal voltages) which can simulate a month of typical use by a buyer in 1 hr of laboratory testing. He is able to take a random sample of 100 bulbs and test them all until failure in less than three months. He does this experiment and records the equivalent typical user month of failure for each bulb. The sample data are given in Table 3.2.

The test engineer wants to use an exponential distribution for bulb failures, based on his past experience. His first step in analyzing this data is, therefore, to decide whether an exponential is a reasonable model.

Table 3.2. Sample Data of Equivalent Month of Bulb Failure.

1	2	2	3	4	5	7	8	9	10
11	13	15	16	17	17	18	18	18	20
20	21	21	24	27	29	30	37	40	40
40	41	46	47	48	52	54	54	55	55
64	65	65	65	67	76	76	79	80	80
82	86	87	89	94	96	100	101	102	104
105	109	109	120	123	141	150	156	156	161
164	167	170	178	181	191	193	206	211	212
214	236	238	240	265	304	317	328	355	363
365	369	389	404	427	435	500	522	547	889

Table 3.3. Frequency Table of Bulb Data.

CELL BOUNDARIES	NUMBER IN CELL
Greater than 0 to 55	40
Greater than 55 to 110	23
Greater than 110 to 165	8
Greater than 165 to 220	10
Greater than 220 to 275	4
Greater than 275 to 330	3
Greater than 330 to 385	4
Greater than 385 to 440	4
Greater than 440 to 495	0
Greater than 495 to 550	3
Greater than 550	1
Total count	100

Using the techniques of Chapter 1, we will construct a histogram for this data. First, however, we note how much larger our intervals will have to be in order to include the last point, because of its distance from the other data points. So we ignore the last point and use $547 - 1 = 546$ instead of the actual range of the data. Dividing this by 10 gives 54.6 or 55 for a cell width. For ease of calculation, we will make the first interval 0 to and including 55. The second interval is greater than 55 to and including 110, and so on. Table 3.3 contains the frequency table for these intervals.

By comparing the shape of the histogram in Figure 3.3 to the exponential PDF shape in Figure 3.2, we see that an exponential model is a reasonable choice for this data. Later in this chapter, we will show how to choose an estimate of λ from the data, and test statistically whether an exponential model with that λ is an acceptable fit.

This example also introduced a very useful concept: that of accelerated testing. How does one verify that testing 1 hr at a given condition is equivalent to 10 or 100 or some other number of typical use condition hours? What is the mathematical basis behind "acceleration" and what data analysis tools are needed? Chapter 7 will deal with this important topic.

THE MEAN TIME TO FAIL

The mean for a life distribution, as defined in Chapter 1, may be thought of as the population average or mean time to fail. In other words, a brand

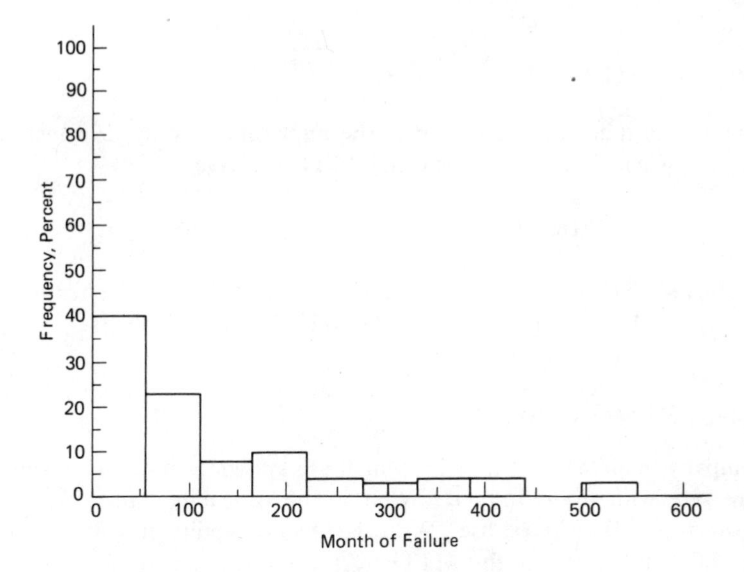

Figure 3.3. Histogram of Bulb Failure Data.

new unit has this expected lifetime until it fails. We abbreviate mean time to fail as MTTF. For the exponential, the definition is

$$\text{MTTF} = \int_0^\infty t\lambda e^{-\lambda t}\, dt.$$

This expression can be integrated by parts to yield

$$\text{MTTF} = \frac{1}{\lambda}.$$

We interpret this result as follows: The MTTF for a population with a constant failure rate λ is the reciprocal of that failure rate or $1/\lambda$.

Even though $1/\lambda$ is the average time of failure, however, it is not also the time when half the population will have failed. This median time to failure, or T_{50}, was described for sample data in Chapter 1. For the entire population, the median is defined to be the point where the CDF function first reaches the value .5. For the exponential we have

$$F(T_{50}) = .5 = 1 - e^{-\lambda T_{50}}.$$

Taking natural logarithms and solving for T_{50} yields

$$T_{50} = \frac{\ln 2}{\lambda} \approx \frac{.693}{\lambda}.$$

This is less than the MTTF, since the numerator is only .693 instead of 1. In fact, when time has reached the MTTF we have

$$F(\text{MTTF}) = 1 - e^{-\lambda/\lambda} = 1 - e^{-1} \approx .632.$$

This shows that approximately 63.2% of an exponential population with failure rate λ has failed by the time the MTTF $1/\lambda$ is reached.

EXAMPLE 3.4 MEAN TIME TO FAIL

A company manufactures resistors which are known to have an exponential failure rate with $\lambda = .15\%/\text{K}$. What is the probability any one resistor will survive 20,000 hr of use? What is the probability it will fail in the next 35,000 hr? What is the MTTF? At what point in time will 10% of these resistors be expected to fail? 50%? When will 63.2% have failed?

SOLUTION

First we convert λ to failures per unit hours of time from %/K using the 10^{-5} conversion factor (if λ had been given in FITs, the factor would be 10^{-9}). Then $\lambda = .15 \times 10^{-5}$ or .0000015.

The probability of surviving 20,000 hr is $R(20{,}000) = e^{-.03}$ or .97. The probability of failing in the next 35,000 hr after surviving to 20,000 hr is a conditional probability with value $[F(55{,}000) - F(20{,}000)]/R(20{,}000)$ or $(.079 - .030)/.97 = .051$.

The MTTF is $1/\lambda$ or $1/.0000015 = 666{,}667$ hr.

We find out when 10% will have failed (known as the tenth percentile) by solving for t in $F(t) = .1$.

$$1 - e^{-.0000015t} = .1$$

$$-.0000015t = \ln .9$$

$$t = \frac{.10536}{.0000015} = 70{,}240 \text{ hr}.$$

Finally, the time when 50% of the population has failed is the median or T_{50} point given by $.693/\lambda = 462{,}000$ hr, and the 63.2% point is the MTTF or 666,667 hr.

In the last example we calculated the probability of failing in 35,000 hr, after surviving 20,000 hr. This turned out to be .051. But the probability of a new resistor failing in its first 35,000 hr, or $F(35{,}000)$, also equals .051. Previous stress time does not seem to make any difference. This property of the exponential distribution is discussed in the next section.

LACK OF MEMORY PROPERTY

The constant failure rate was one of the characteristic properties of the exponential: closely related is another key property—the exponential "lack of memory." A component following an exponential life distribution does not "remember" how long it has been operating. The probability it will fail in the next hour of operation is the same if it were new, one month old, or several years old. It does not age or wear out or degrade with time or use. Failure is a chance happening, always at the same constant rate, unrelated to accumulated power-on hours.

The equation that describes this property says that the conditional probability of failure in some interval of time of length h, given survival up to the start of that interval, is the same as the probability of a new unit failing in its first h hr.

$$P(\text{fail in next } h | \text{survive } t) = P(\text{new unit fails in } h).$$

In terms of the CDF this becomes

$$\frac{F(t+h) - F(t)}{1 - F(t)} = F(h).$$

This equation holds if $F(t)$ is the exponential CDF as the reader can easily verify. It can also be shown that only the exponential has this property for all t and h (see Feller, 1968, page 459).

The implications of this concept from a testing point of view are highly significant. We gain equivalent information from testing 10 units for 20,000 hr or from testing 1000 units for 20 hr, (or even 1 unit for 200,000 hr). If a unit fails on test, we have the option of repairing or replacing it and continuing the test without worrying about the fact that some of the test units have a different "age" than other units.

If engineering judgment says that the above testing equivalences seem wrong in a particular case, then we are really saying that we do not believe the exponential distribution applies. However, when the exponential can be used as a reasonable model for the data and the type of item on test, these advantages in test design apply.

Another consequence of the "lack of memory" property is that the renewal rate (defined in Chapter 2), and the failure rate, (as well as the average failure rate) are all equal and have value λ. This fact takes away concern over the issue discussed in Chapter 2: namely, which rate is the best to use in a particular situation? We can even define an expected time between failures for a repairable, exponentially distributed unit. This mean time between failures, or MTBF, is again λ.

AREAS OF APPLICATION

If we feel that a unit under test has no significant wearout mechanisms, at least for its intended application life, and either we do not expect many early defect failures or we intend to separate these out and treat them separately, then the exponential is a good initial choice for a life distribution model.

Note the words "no significant wearout mechanisms, at least for its intended application life." Even though we know, in every real world case we can imagine, some kind of wearout eventually takes place, we can ignore this if we feel it is not a practical concern. Having a nearly constant failure rate over the region of time we are interested in (and confining our testing time to this region, or its equivalence under test acceleration) is all we need to use the exponential.

Another application of the exponential is in modeling the long, flat portion of the bathtub curve. Since use failure data on many systems, subassemblies, or even individual components, has a nearly constant failure rate for most of the product life, exponential methods of data analysis can be used successfully.

In fact, if we specify product performance over an interval of time using an average failure rate, and we are not particularly concerned with how the failures spread out over that interval, then we can use an exponential assumption and the exponential confidence bound factors to be described later in this chapter. This feature extends the usefulness of the exponential to many cases where it would not be the correct theoretical model. Just as any smooth curve can be approximated piece-wise by straight lines to any degree of accuracy required, we can consider a changing failure rate curve to be constructed of many piece-wise constant exponential portions. Then we can analyze data from within any one piece, or interval, as if it were exponential.

Later in this chapter, we will see that the process of planning experiments where the exponential assumption applies is very straightforward. For example, sample sizes and confidence levels and precision can be decided in advance. It will be much more difficult, or even impossible, to do similar exact planning for other life distribution models. For this reason, the exponential is also useful as a trial model in the experimental planning stage, even if we do not expect it actually to apply later on.

One must be careful, however, when setting up an experiment designed to accelerate a known wearout mechanism. If the purpose is to determine whether wearout will, indeed, start well beyond useful product life, then we are clearly interested in the nonflat rising portion of the bathtub curve. Exponential analysis methods would not apply, and preplanning should be based on the actual life distribution model, if at all possible.

ESTIMATION OF λ

When data come from an exponential distribution, there is only one parameter, λ, to estimate. The best estimate for complete or censored samples is

$$\hat{\lambda} = \frac{\text{number of failures}}{\text{total unit test hours}}.$$

The denominator is obtained by adding up all the operation hours on test of all the units tested, both those that failed and those that completed test without failing.

For a complete sample (everything fails and exact times of failure are recorded), this expression reduces to the reciprocal of the sample mean. Thus we have $\hat{\lambda} = 1/(\text{sample mean time to failure})$, just as we had $\lambda = 1/\text{MTTF}$.

For censored Type I data (fixed test time), with r failures out of n on test

$$\hat{\lambda} = \frac{r}{\sum_{i=1}^{r} t_i + (n-r)T}.$$

T is the pre-fixed end of test time and $t_1, t_2, t_3, \ldots, t_r$ are the exact failure times of the r units that fail before the test ends.

If the test is censored Type II (ends at rth failure time), the same rule yields

$$\hat{\lambda} = \frac{r}{\sum_{i=1}^{r} t_i + (n-r)t_r}.$$

If new units are put on test to replace failed units, or to increase the sample size part way through the testing, then we have multicensored data.

Applying the general form for $\hat{\lambda}$, the denominator is the sum of each test unit's time on test.

When we have readout data, we can no longer exactly calculate the numerator in order to estimate λ. In this case, the graphical methods described in Chapter 6 can be used (they also apply, as an alternate approach, when exact times are available). More precise techniques, based on the method of maximum likelihood estimation, will be described in Chapter 4.

We can also apply the simple procedure described in this section to readout data, often with little loss of accuracy. For example, if many units are on test with only a few failures, the error in assuming all failures occur in the middle of the readout interval will be negligible. With this assumption, the simple confidence bound procedures described later in this chapter can be used.

It should be noted that when exact times are available, the estimation rule described here yields estimates that are also maximum likelihood estimates (see definition in Chapter 4 and Example 4.4).

EXAMPLE 3.5 FAILURE RATE AND MTTF

Returning to the light bulb data of Example 3.3 and Table 3.2, the estimate of λ is $\hat{\lambda} = 100/13{,}563 = .0074$. The MTTF estimate $1/\hat{\lambda}$ is 135.6 months.

On the other hand, if we had used the summarized readout data version of bulb failure months given in Table 3.3, we would have calculated $\hat{\lambda} = 100/13{,}505 = .0074$ and MTTF = 135.1. For this calculation, the denominator would be obtained by multiplying the 40 failures in the first interval by the middle of the interval, or 27.5, and adding 23×82.5 for the second interval, and so on. For the one fail time after 550 months, we pretend our 55-month intervals go on to 935 months, where the last fail is recorded. Placing this fail in the center of that last interval (880 to 935) at 907.5 months, gives the total test time of 13,505 months (the reader should try this calculation).

This example shows that using the failures divided by total test time formula on readout data is not likely to cause much error.

One immediate use we can make of this λ estimate is to use it to calculate values of the PDF to compare to our histogram in Figure 3.3. In other words, the histogram is supposed to have the same shape as $f(t)$, which we are estimating by $f(t) = .0074\, e^{-.0074t}$.

Before graphing $f(t)$, however, and putting it on the same chart with the histogram, we have to adjust scales. The total area under the $f(t)$ curve is always 1. The histogram we plotted, because the intervals have width 55 and the height units are in percent, has an area of 100×55 or 5500 (if we put one more box of height 1 between 880 and 935). In Figure 3.4, we

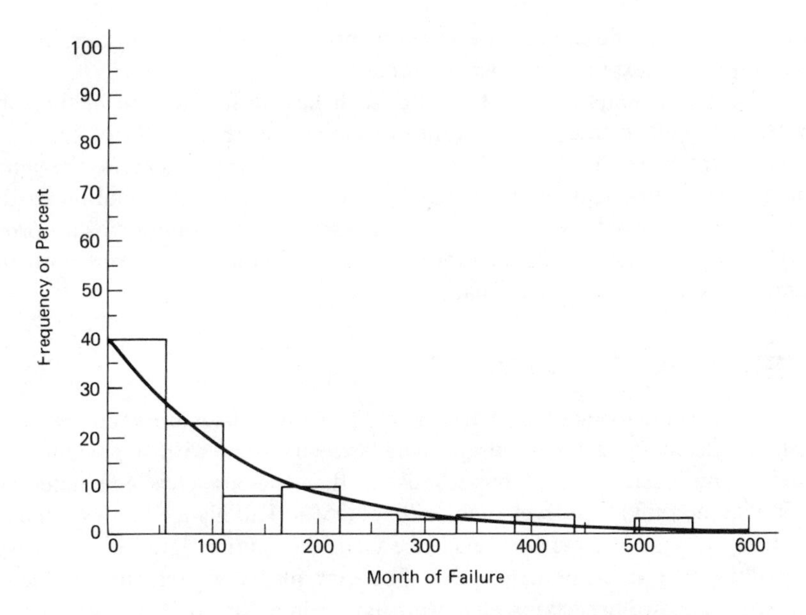

Figure 3.4. Light Bulb Data Histogram Compared to $f(t)$ Shape.

plot $5500 \times f(t)$ along with the light bulb data histogram. This gives us a direct shape comparison to show that an exponential model applies. In the section after the next, we will test this fit with the chi-square goodness of fit test.

EXPONENTIAL DISTRIBUTION CLOSURE PROPERTY

The exponential distribution possesses a convenient closure property that applies to assemblies or systems made up of exponentially distributed components. If the system fails when the first component fails, and all the components operate independently, then the system life distribution is also exponential. The failure rate parameter for the system is equal to the sum of the failure rate parameters of the components.

A system model where n components operate independently and the system fails with the first component failure is called a series model. This model is discussed in Chapter 8, where it is shown that the system failure rate function $h_s(t)$ is the sum of the n component failure rate functions $h_1(t), h_2(t), \ldots, h_n(t)$. When the components have exponential lifetimes with parameters $\lambda_1, \lambda_2, \ldots, \lambda_n$, then the system has a constant failure rate equal to

$$\lambda_s = \sum_{i=1}^{n} \lambda_i .$$

This establishes the exponential closure property, since a constant failure rate implies an exponential distribution.

If the components are all the same, each having failure rate λ, then the system has failure rate $n\lambda$ and a mean time to failure (MTTF) of $1/n\lambda$.

This exponential closure property is almost unique. In general, the minimum of n components will not have the same type life distribution as the components themselves. (There is another case where a similar closure property applies—when all the components have Weibull distributions with the same shape parameter—see Chapter 4.)

TESTING GOODNESS OF FIT

There is a standard statistical test, known as the chi-square goodness of fit test, for deciding whether sample data are consistent with a preconceived model. Our basic, or null hypothesis, is that the model is adequate. We then pick a confidence level such as 90 or 95%. The higher the level means we want very strong negative evidence from the sample data before we will be willing to reject our hypothesis. For example, if we set our confidence level at 90%, we are saying that we need sample data that occur less than 10% of the time if our model is correct, before we will reject it. If we set our confidence level to 95%, we only reject on data that occur less than 5% of the time if the model is correct.

In other words, what we really control by picking a confidence level is the probability of making a mistake by rejecting a good model. This probability of error is known as the Type I error, and has value less than .01 times 100 minus the confidence level. (A more detailed treatment of the probabilities of making decision errors when sampling is given in Chapter 9.)

In the goodness of fit test, we go through a set procedure, about to be described, and calculate a number. If our chosen model is correct, this random quantity we have calculated (random because its value depends on the random sample data) will have an approximate chi-square distribution. This is a well-known distribution which has percentile values that are tabled in the Appendix. If the calculated number turns out so large that the tables indicate it is a highly unlikely value, then we reject the model. Any value higher than the confidence level percentile is defined to be "highly unlikely" enough.

The above discussion may sound somewhat abstract and academic. However, it does have a very serious consequence that is not always understood or appreciated by those who use statistical hypothesis testing methods. When we perform a statistical test and end up not rejecting the assumed hypothesis, we have not proved it is correct. The terminology "accept" the hypothesis and "90% confidence level" may mislead us into thinking we are confident our hypothesis is true. The statistical test gives us no such confidence. We started with what we thought was a reasonable model, and all we have shown is that the sample data have not changed our mind.

When we reject, on the other hand, we are making a strong statement of belief. We are saying that the sample evidence was so overwhelmingly against the model we had chosen that we have to reconsider our choice.

Readers who desire more background on the theory of hypothesis testing should consult a basic statistics textbook such as Mendenhall and Scheafer (1983). We now turn to how to test whether sample data are consistent with a distributional assumption such as exponential.

The steps are as follows:

1. Group the data, if necessary, into intervals as if preparing to plot a histogram. The intervals need not be equal, however. Form a table with the interval and the observed frequency, as in Table 3.3 for the light bulb data.
2. If the model is a completely specified distribution (i.e., no unknown parameters), go on to step 3. Otherwise estimate all the unknown parameters, using the method of maximum likelihood. In the case of the exponential, this means estimate λ using the method previously described. At this point, the model is completely specified.
3. Use the CDF implied by the model to calculate the probability of failure in each of the intervals listed in the table. If the interval is (I_1,I_2), this probability is $F(I_2) - F(I_1)$. Add a column with these probabilities to the table.
4. Multiply the probabilities just calculated (which should total to 1) by the sample size on test. This calculation results in an expected number of failures for each interval. Add a column of these numbers to the table.
5. If any intervals have less than five expected failures (not actual) as just calculated, these intervals must be combined with other intervals until every number in the expected column is 5 or greater.
6. Now calculate [(actual failures − expected failures)^2 divided by expected failures] for each of the remaining intervals on the table. Add these numbers together. This is the χ^2 test statistic.
7. Take 1 less than the number of separate intervals in the table (or the number of terms added together in step 6 less 1) and subtract the number of parameters estimated in step 2. The number obtained is the degrees of freedom of the χ^2 test statistic.
8. Compare the χ^2 statistic to the table values in the Appendix for the calculated degrees of freedom. Look to see if the test statistic is higher

than the value in the table for the chosen confidence level. If it is higher, reject the model. If not, continue to use the model, but note just how close you were to rejecting it.

The grouping in step 5 is necessary because the test statistic calculated in step 6 has only an approximate chi-square distribution. As long as each interval has at least five expected failures, the approximation is good enough to be useful; for under five expected failures the chi-square approximation may be inappropriate.

EXAMPLE 3.6 CHI-SQUARE GOODNESS OF FIT

Table 3.4 shows the worksheet obtained from applying the first four steps in calculating a chi-square test statistic to the light bulb failure month data in Table 3.3. The estimate of $\hat{\lambda} = .0074$ obtained in Example 3.5 has been used. The probability value for the 55 to 110 interval will be calculated in detail to illustrate the procedure.

$$\begin{aligned} P(\text{fail in 50 to 110 interval}) &= F(110) - F(55) \\ &= (1 - e^{-.0074\times 110}) - (1 - e^{-.0074\times 55}) \\ &= e^{-.407} - e^{-.814} \\ &= .666 - .443 = .223. \end{aligned}$$

Table 3.4. Chi-Square Goodness of Fit Worksheet for the Light Bulb Data.

INTERVAL (I_1, I_2)	ACTUAL FAILS	P(FAIL IN INTERVAL) $F(I_2) - F(I_1)$	EXPECTED FAILURES
0 to 50	40	.334	33.4
50 to 110	23	.223	22.3
110 to 165	8	.148	14.8
165 to 220	10	.099	9.9
220 to 275	4	.066	6.6
275 to 330	3	.044	4.4
330 to 385	4	.029	2.9
385 to 440	4	.019	1.9
440 to 495	0	.013	1.3
495 to 550	3	.009	.9
>550	1	.017	1.7

In order to have 5 or more failures in every interval, we can combine 275 to 330 and 330 to 385 to have an expected of 7.3 versus an actual of 7. We also have to combine the last three intervals into a single "greater than 385" interval with $1.9 + 1.3 + .9 + 1.7 = 5.8$ expected and $4 + 0 + 3 + 1 = 8$ observed.

The calculation described in step 6 becomes

$$\begin{aligned}
&\frac{(40-33.4)^2}{33.4} + \frac{(23-22.3)^2}{22.3} \\
&+ \frac{(8-14.8)^2}{14.8} + \frac{(10-9.9)^2}{9.9} \\
&+ \frac{(4-6.6)^2}{6.6} + \frac{(7-7.3)^2}{7.3} \\
&+ \frac{(8-5.8)^2}{5.8} \\
&= 1.30 + .02 + 3.12 + 0 + 1.02 + .01 + .83 \\
&= 6.3.
\end{aligned}$$

This test statistic was a sum of seven terms and the rule for degrees of freedom given in step 7 says to subtract 1 and subtract another 1 because λ was estimated from the data. This give 5 for the degrees of freedom. The chi-square table in the Appendix has 11.1 for a 95% point and 9.24 for a 90% point. The test statistic is only 6.3, which is a little higher than the 70% point. That means there is no reason to reject an exponential model for the population this data come from.

CONFIDENCE BOUNDS FOR λ AND THE MTTF

The estimate $\hat{\lambda}$ = number of failures/total unit test hours is a single number or point estimate of λ. Even though it may be the "best" estimate we can come up with, by itself it gives no measure of precision or risk. Can the true λ be as high as $10 \times \hat{\lambda}$? Or are we confident it is no worse than $1.2 \times \hat{\lambda}$? And how much better than $\hat{\lambda}$ might the true λ actually be?

These are very important questions if critical decisions must be made which depend on the true value of λ. No presentation of the test results and a calculation of $\hat{\lambda}$ are complete without including an interval around $\hat{\lambda}$ which has a high degree of confidence of enclosing the true value of λ.

This kind of interval is called a confidence interval. A 90% confidence interval means that if the same experiment were repeated many times, and the same method was used over and over again to construct an interval for

λ, 90% of these intervals would contain the true λ. For the one time we actually do the experiment, our interval either does or does not contain λ. But if it is a 90% confidence interval, then we would give 9 to 1 odds in its favor.

For Type I (time censored) or Type II (rth failure censored) data, factors based on the chi-square distribution can be derived and used as multipliers to obtain the upper and lower ends of any confidence intervals. The remarkable thing about these factors is they depend only on the number of failures observed during the test.

Suppose we want a $100 \times (1 - \alpha)$ confidence interval for λ. For example, $\alpha = .1$ corresponds to a 90% interval. We will calculate a lower $100 \times \alpha/2$ bound for λ and an upper $100 \times (1 - \alpha/2)$ bound for λ. These two numbers give the desired confidence interval. When α is .1, this becomes a lower 5% bound and an upper 95% bound, having between them 90% chance of containing λ. The lower end of this interval, denoted by λ_5, is a 95% single-sided lower bound for λ. The upper end, or λ_{95}, is a single-sided 95% upper bound for λ.

The notation here can be confusing at first and the reader should work out several examples until it is clear. As an illustration, suppose we want a 95% interval for λ. Then $\alpha = .05$ and $\alpha/2 = .025$. The interval will be $(\lambda_{2.5}, \lambda_{97.5})$. The lower end is a 97.5% lower bound for λ and the upper end is a 97.5% upper bound.

Now that we have defined our notation, how do we calculate these lower and upper bounds? When we have Type II censored data, it can be shown that the lower $100 \times \alpha/2$ percentile of the chi-square distribution with $2r$ degrees of freedom (r is the number of failures), divided by $2r$, is a factor we can multiply $\hat{\lambda}$ by to get $\lambda_{100 \times \alpha/2}$. Similarly, the upper $100 \times (1 - \alpha/2)$ percentile of the same chi-square, divided by $2r$, is a factor we can multiply $\hat{\lambda}$ by to get $\lambda_{100 \times (1-\alpha/2)}$.

For the more common Type I censored data, intervals using the above factors are approximately correct. Exact intervals can be calculated only if failed units are replaced immediately, during the course of the test. In this case, the lower bound factor is exactly as above, while the upper factor uses the $100 \times (1 - \alpha/2)$ percentile of a chi-square with $2 \times (r + 1)$ degrees of freedom, still divided by $2r$. Since this chi-square factor produces a slightly more conservative upper bound, we recommend using it for Type I censoring.

The factors described above depend only on the number of failures r. We will denote the factor that generates a one-sided upper $100 \times (1 - \alpha)$ bound by $k_{r;1-\alpha}$. The key equation is

$$k_{r;1-\alpha} \times \hat{\lambda} = \lambda_{100 \times (1-\alpha)}$$

Table 3.5. Factors for 1-Sided Exponential Upper Bound (Type I Time Censoring).

NO. FAILS	60%	80%	90%	95%	97.5%	99%	99.9%	NO. FAILS	60%	80%	90%	95%	97.5%	99%	99.9%
1	2.02	2.99	3.89	4.74	5.57	6.64	9.25	42	1.05	1.15	1.23	1.29	1.35	1.42	1.58
2	1.55	2.14	2.66	3.15	3.61	4.21	5.63	43	1.05	1.15	1.23	1.29	1.35	1.42	1.57
3	1.39	1.84	2.23	2.58	2.92	3.35	4.33	44	1.05	1.15	1.22	1.29	1.34	1.41	1.57
4	1.31	1.68	2.00	2.29	2.56	2.90	3.67	45	1.05	1.15	1.22	1.28	1.34	1.41	1.55
5	1.26	1.58	1.86	2.10	2.33	2.62	3.30	46	1.05	1.14	1.22	1.28	1.33	1.40	1.55
6	1.22	1.51	1.76	1.97	2.18	2.43	2.99	47	1.05	1.14	1.21	1.28	1.33	1.40	1.54
7	1.20	1.46	1.68	1.88	2.06	2.29	2.79	48	1.05	1.14	1.21	1.27	1.33	1.39	1.53
8	1.18	1.42	1.62	1.80	1.97	2.18	2.64	49	1.05	1.14	1.21	1.27	1.32	1.39	1.53
9	1.16	1.39	1.58	1.74	1.90	2.09	2.53	50	1.05	1.14	1.21	1.27	1.32	1.38	1.52
10	1.15	1.37	1.54	1.70	1.84	2.02	2.41	55	1.05	1.13	1.20	1.25	1.30	1.36	1.50
11	1.14	1.34	1.51	1.66	1.79	1.95	2.32	60	1.04	1.12	1.19	1.24	1.29	1.34	1.47
12	1.13	1.33	1.48	1.62	1.75	1.90	2.25	65	1.04	1.12	1.18	1.23	1.27	1.33	1.44
13	1.12	1.31	1.46	1.59	1.71	1.86	2.19	70	1.04	1.11	1.17	1.22	1.26	1.32	1.43
14	1.12	1.29	1.44	1.56	1.68	1.82	2.14	75	1.04	1.11	1.16	1.21	1.25	1.30	1.41
15	1.11	1.28	1.42	1.54	1.65	1.78	2.08	80	1.04	1.11	1.16	1.20	1.24	1.29	1.40
16	1.11	1.27	1.40	1.52	1.62	1.75	2.04	85	1.04	1.10	1.15	1.20	1.24	1.28	1.38
17	1.10	1.26	1.39	1.50	1.60	1.72	2.00	90	1.03	1.10	1.15	1.19	1.23	1.27	1.37
18	1.10	1.25	1.38	1.48	1.58	1.70	1.96	95	1.03	1.10	1.14	1.19	1.22	1.27	1.36
19	1.09	1.24	1.36	1.47	1.56	1.68	1.94	100	1.03	1.09	1.14	1.18	1.22	1.26	1.35
20	1.09	1.24	1.35	1.45	1.54	1.65	1.90	110	1.03	1.09	1.13	1.17	1.21	1.25	1.33
21	1.09	1.23	1.34	1.44	1.53	1.64	1.87	120	1.03	1.08	1.13	1.16	1.20	1.23	1.32
22	1.09	1.22	1.33	1.43	1.51	1.62	1.85	130	1.03	1.08	1.12	1.16	1.19	1.22	1.30
23	1.08	1.22	1.32	1.42	1.50	1.60	1.83	140	1.03	1.08	1.12	1.15	1.18	1.21	1.29
24	1.08	1.21	1.32	1.41	1.49	1.59	1.81	150	1.03	1.07	1.11	1.15	1.17	1.21	1.28
25	1.08	1.21	1.31	1.40	1.48	1.57	1.79	160	1.02	1.07	1.11	1.14	1.17	1.20	1.27
26	1.08	1.20	1.30	1.39	1.47	1.56	1.77	170	1.02	1.07	1.11	1.14	1.16	1.19	1.26
27	1.07	1.20	1.30	1.38	1.46	1.55	1.75	180	1.02	1.07	1.10	1.13	1.16	1.19	1.25
28	1.07	1.19	1.29	1.37	1.45	1.54	1.73	190	1.02	1.07	1.10	1.13	1.15	1.18	1.24
29	1.07	1.19	1.28	1.36	1.44	1.52	1.71	200	1.02	1.06	1.10	1.12	1.15	1.18	1.24
30	1.07	1.19	1.28	1.36	1.43	1.51	1.71	250	1.02	1.06	1.09	1.11	1.13	1.16	1.21
31	1.07	1.18	1.27	1.35	1.42	1.50	1.69	300	1.02	1.05	1.08	1.10	1.12	1.14	1.19
32	1.07	1.18	1.27	1.34	1.41	1.49	1.68	400	1.02	1.04	1.07	1.09	1.10	1.12	1.16
33	1.07	1.18	1.26	1.34	1.40	1.49	1.66	500	1.01	1.04	1.06	1.08	1.09	1.11	1.15
34	1.06	1.17	1.26	1.33	1.40	1.48	1.66	600	1.01	1.04	1.05	1.07	1.08	1.10	1.13
35	1.06	1.17	1.25	1.33	1.39	1.47	1.64	700	1.01	1.03	1.05	1.06	1.08	1.09	1.12
36	1.06	1.17	1.25	1.32	1.38	1.46	1.63	950	1.01	1.03	1.04	1.05	1.07	1.08	1.10
37	1.06	1.16	1.24	1.31	1.38	1.45	1.62	1500	1.01	1.02	1.03	1.04	1.05	1.06	1.08
38	1.06	1.16	1.24	1.31	1.37	1.45	1.61	3000	1.01	1.01	1.02	1.03	1.04	1.04	1.06
39	1.06	1.16	1.24	1.31	1.37	1.44	1.60	5000	1.00	1.01	1.02	1.02	1.03	1.03	1.04
40	1.06	1.16	1.23	1.30	1.36	1.43	1.59	20000	1.00	1.01	1.01	1.01	1.01	1.02	1.02
41	1.06	1.15	1.23	1.30	1.36	1.43	1.58								

1. Best F/R estimate: $\hat{\lambda} = \frac{\text{no. fails}}{\text{total POH}} \times 10^5(\%/K)$. POH = Power on Hours or total unit test hours.
2. Obtain upper bound at confidence level desired by multiplying by appropriate factor from above table.
3. Mean time to fail estimate (MTTF) $= \frac{1}{.01 \times \hat{\lambda}}$ in KPOH. Lower bound $= \frac{1}{.01 \times \hat{\lambda}_{\text{upper bound}}}$ in KPOH. KPOH = Power on Hours/1000.
4. For Type II censoring at the r-th failure time, use the factor above corresponding to $r - 1$ fails, multiplied by $\frac{(r-1)}{r}$. For $r = 1$, use the constants given in Table 3.7, divided by the total POH instead of divided by nT.

Table 3.6. Factors for 1-Sided Exponential Lower Bound (I and II Censoring).

NO. FAILS	60%	80%	90%	95%	97.5%	99%	99.9%	NO. FAILS	60%	80%	90%	95%	97.5%	99%	99.9%
1	.51	.22	.11	.05	.03	.01	.00	42	.95	.87	.81	.76	.72	.68	.59
2	.69	.41	.27	.18	.12	.07	.02	43	.95	.87	.81	.76	.72	.68	.59
3	.76	.51	.37	.27	.21	.15	.06	44	.95	.87	.81	.77	.73	.68	.60
4	.80	.57	.44	.34	.27	.21	.11	45	.96	.87	.81	.77	.73	.69	.60
5	.83	.62	.49	.39	.32	.26	.15	46	.96	.87	.82	.77	.73	.69	.61
6	.85	.65	.53	.44	.37	.30	.19	47	.96	.88	.82	.77	.73	.69	.61
7	.86	.68	.56	.47	.40	.33	.22	48	.96	.88	.82	.77	.74	.70	.61
8	.87	.70	.58	.50	.43	.36	.25	49	.96	.88	.82	.78	.74	.70	.62
9	.88	.71	.60	.52	.46	.39	.27	50	.96	.88	.82	.78	.74	.70	.62
10	.89	.73	.62	.54	.48	.41	.30	55	.96	.89	.83	.79	.75	.71	.63
11	.90	.74	.64	.56	.50	.43	.32	60	.96	.89	.84	.80	.76	.72	.65
12	.90	.75	.65	.58	.52	.45	.34	65	.96	.89	.84	.81	.77	.73	.66
13	.91	.76	.67	.59	.53	.47	.35	70	.97	.90	.85	.81	.78	.74	.67
14	.91	.77	.68	.60	.55	.48	.37	75	.97	.90	.85	.82	.79	.75	.68
15	.91	.78	.69	.62	.56	.50	.38	80	.97	.90	.86	.82	.79	.76	.69
16	.92	.79	.70	.63	.57	.51	.40	85	.97	.91	.86	.83	.80	.77	.70
17	.92	.79	.70	.64	.58	.52	.41	90	.97	.91	.87	.83	.80	.77	.70
18	.92	.80	.71	.65	.59	.53	.42	95	.97	.91	.87	.84	.81	.78	.71
19	.93	.80	.72	.65	.60	.54	.44	100	.97	.92	.87	.84	.81	.78	.72
20	.93	.81	.73	.66	.61	.55	.45	110	.97	.92	.88	.85	.82	.79	.73
21	.93	.81	.73	.67	.62	.56	.46	120	.97	.92	.89	.85	.83	.80	.74
22	.93	.82	.74	.68	.63	.57	.47	130	.98	.93	.89	.86	.84	.81	.75
23	.93	.82	.74	.68	.63	.58	.48	140	.98	.93	.89	.87	.84	.81	.76
24	.94	.83	.75	.69	.64	.59	.48	150	.98	.93	.90	.87	.85	.82	.77
25	.94	.83	.75	.70	.65	.59	.49	160	.98	.93	.90	.87	.85	.83	.77
26	.94	.83	.76	.70	.65	.60	.50	170	.98	.94	.90	.88	.85	.83	.78
27	.94	.84	.76	.71	.66	.61	.51	180	.98	.94	.91	.88	.86	.83	.79
28	.94	.84	.77	.71	.66	.61	.52	190	.98	.94	.91	.88	.86	.84	.79
29	.94	.84	.77	.72	.67	.62	.52	200	.98	.94	.91	.89	.87	.84	.80
30	.94	.84	.77	.72	.67	.62	.53	250	.98	.95	.92	.90	.88	.86	.82
31	.94	.85	.78	.72	.68	.63	.53	300	.98	.95	.93	.91	.89	.87	.83
32	.95	.85	.78	.73	.68	.63	.54	400	.99	.96	.94	.92	.90	.89	.85
33	.95	.85	.78	.73	.69	.64	.55	500	.99	.96	.94	.93	.91	.90	.87
34	.95	.85	.79	.74	.69	.64	.55	700	.99	.97	.95	.94	.93	.91	.89
35	.95	.86	.79	.74	.70	.65	.55	900	.99	.97	.96	.95	.94	.92	.90
36	.95	.86	.79	.74	.70	.65	.56	2000	.99	.98	.97	.96	.96	.92	.93
37	.95	.86	.80	.75	.70	.66	.57	3000	1.00	.98	.98	.97	.96	.96	.94
38	.95	.86	.80	.75	.71	.66	.57	5000	1.00	.99	.98	.98	.97	.97	.96
39	.95	.86	.80	.75	.71	.67	.58	20000	1.00	.99	.99	.99	.99	.98	.98
40	.95	.86	.80	.75	.71	.67	.58	50000	1.00	1.00	.99	.99	.99	.99	.99
41	.95	.87	.81	.76	.72	.67	.59								

1. Best F/R estimate $\hat{\lambda}\ \frac{\text{no. fails}}{\text{total POH}} \times 10^5\ (\%/K)$.

for a Type I censored experiment with r failures. The factors are given in two tables. Table 3.5 has upper bound k factors. Table 3.6 has lower bound factors. Use of these tables is shown in Example 3.8. Note: since the tables give one-sided confidence bounds factors, in order to obtain, for example, a 90% two-sided interval, one would use the factors corresponding to the 95% upper and lower columns.

Adjustment For Type II Censoring

If the data are Type II censored, the lower bound factors from Table 3.6 still apply. To get the proper upper bound factor from Table 3.5 the following trick must be used: look along the row for $r - 1$ failures, instead of r. Find the k factor at the desired confidence and adjust it by multiplying by $(r - 1)/r$. This adjusted factor is the correct Type II upper bound multiplier. (See Example 3.8.) For $r = 1$, use the constants given in Table 3.7, divided by the total POH instead of divided by nT.

THE CASE OF ZERO FAILURES

When a test ends after T hr with none of the n test units having failed, the point estimate previously defined is zero. This is not a realistic estimate, as it does not even take into account the number on test. An upper $100 \times (1 - \alpha)$ confidence limit for λ is given by

$$\lambda_{100\times(1-\alpha)} = \frac{\chi^2_{2;100\times(1-\alpha)}}{2nT},$$

where $\chi^2_{2;100\times(1-\alpha)}$ is the upper $100 \times (1 - \alpha)$ percentile of the chi-square distribution with 2 degrees of freedom.

The 50% zero failures estimate is often used as a point estimate for λ. This should be interpreted very carefully. It is a value of λ which makes the likelihood of obtaining zero failures in the given experiment similar to the chance of getting a head when flipping a coin. We are not really 50% confident of anything; we have just picked a λ that will produce zero failures 50% of the time.

Table 3.7 gives zero fail $\lambda_{100\times(1-\alpha)}$ formulas for several percentiles. A general formula that can be used to derive values for any percentile is $\lambda_{100\times(1-\alpha)} = (-\ln\alpha)/nT$.

EXAMPLE 3.7 ZERO FAILURES ESTIMATION

Two hundred samples from a population of units believed to have a constant failure rate are put on test for 2000 hr. At that time, having observed no

Table 3.7. Exponential Zero Fail Estimates.

PERCENTILE	ESTIMATE
50	$.6931/nT$
60	$.9163/nT$
80	$1.6094/nT$
90	$2.3026/nT$
95	$2.9957/nT$
97.5	$3.6889/nT$
99	$4.6052/nT$

failures, the experimenter puts an additional 200 units on test. Three thousand hours later there are still no failures and all the units are removed from test. What is a 50% estimate of the failure rate? A 95% upper bound? Use the general formula to calculate a 70% upper bound on the true failure rate.

SOLUTION

The total unit test hours are 200 × 5000 plus 200 × 3000 or 1,600,000 hr. The 50% estimate is .6931/1,600,000 which is 433 FITs. The 95% failure rate upper bound is 2.9957/1,600,000 or 1872 FITs. For a 70% upper bound, $\alpha = .3$ and the general formula yields $\lambda_{100\times(1-\alpha)} = (-\ln .3)/1{,}600{,}000 =$ 752 FITs.

So far, all upper and lower bounds have been for the failure rate and not the MTTF. But since the MTTF is $1/\lambda$, we can work with $\hat{\lambda}$ and bounds on λ, and then take reciprocals to convert to MTTF estimates with bounds. Note that the reciprocal of the upper bound for λ becomes the lower bound for the MTTF.

EXAMPLE 3.8 CONFIDENCE BOUNDS ON MTTF

Two hundred units were tested 5000 hr with 4 failures. Give 95% upper and lower bounds on the failure rate and the MTTF. What difference would it make if the test were designed to end at the fourth failure?

SOLUTION

The $\hat{\lambda}$ estimate is 4/1,000,000. From the tables with $r = 4$, we find the upper bound factor of 2.29 and the lower bound factor of .34. After multiplying $\hat{\lambda}$ by these factors we obtain $\lambda_5 = 1.36 \times 10^{-6} = .136\%/K$ and $\lambda_{95} = 9.16 \times 10^{-6} = .916\%/K$. These taken together form a 90% confidence interval for λ. The 95% lower bound on the MTTF is $1/\lambda_{95} = 109{,}170$ hr. The 95% upper bound on the MTTF is $1/\lambda_5 = 735.294$.

If the data were Type II censored at the fourth fail, we would have the same $\hat{\lambda}$ and λ_5. The factor to obtain λ_{95} would be derived by taking the factor 2.58 from $r = 3$ in Table 3.5 and multiplying it by ¾. This gives a result of 1.935. Thus $\lambda_{95} = 1.935 \times 4 \times 10^{-6}$ or $7.74 \times 10^{-6} = .774\%/K$.

PLANNING EXPERIMENTS USING THE EXPONENTIAL DISTRIBUTION

Proper experimental planning, or good experimental design, is acknowledged to be one of the most important ingredients of successful experimentation. Unfortunately, it is often neglected in the case of reliability testing and modeling. This neglect is because the complexity of censored data, as well as the difficult forms of many of the life distributions used in reliability analysis, present problems in experimental design which are mostly unsolved.

This planning difficulty is not present when using the exponential distribution. Here we can, and should, give early consideration to the sample sizes and test durations of the experiment. If we carefully state our objectives, we can plan the right experiment to achieve them. In most cases, only a simple look-up of the right $k_{r,\,1-\alpha}$ factor from the preceding section is necessary.

Case I. How Many Units Should Be Put on Test

The following items must be specified before a sample size can be chosen: a failure rate or MTTF objective; a confidence level for assuring we meet this objective; a test duration; and a somewhat arbitrary number of failures we decide in advance we want to allow on the test and still meet the objective.

With all these items specified, we know all the terms in the equation

$$\frac{r}{n \times T} \times k_{r;\,1-\alpha} = \lambda_{\text{obj}}$$

except n (the denominator is only an approximation of the total unit test hours for planning purposes). Solving this equation for n gives the required sample size.

EXAMPLE 3.9 CHOOSING SAMPLE SIZES

We wish to be 90% confident of meeting a .2%/K specification (MTTF = 500K hr). We can run a test for 5000 hr, and we agree to allow up to 5 failures and still pass the product. What sample size is needed?

SOLUTION

The $k_{5;.90}$ factor from Table 3.5 is 1.86. The basic equation is

$$(5/(5000 \times n)) \times 1.86 = .000002$$
$$n = 930.$$

Variation: How Long Must the Test Run?

Here, we have a fixed number of test units and have specified the failure rate objective, confidence level, and number of allowable failures. The only unknown remaining in the basic equation is the test time.

EXAMPLE 3.10 CHOOSING TEST TIMES

We have 100 units to test and we want to be 95% confident that the MTTF is greater than 20,000 hr. We will allow up to 10 failures. How long must the test run?

SOLUTION

The $k_{10;.95}$ factor is 1.7. Solving for T in

$$(10/(100 \times T)) \times 1.7 = .00005$$

results in a test time of 3400 hr.

Case II. With a Test Plan in Place, How Many Fails Can Be Allowed

It is good practice to clearly state the pass/fail criteria of a test in advance and have all interested parties agree on it. After the experiment is run, it is much more difficult to obtain such agreements.

As before, a failure rate or MTTF objective must be stated, along with the confidence level. The test plan will fix the sample size and test length. From the basic equation, we can solve for $r \times k_{r;1-\alpha}$. Since r is the only unknown, we can not pick the $k_{r;\,1-\alpha}$ from the table. We can, however, move down the $1-\alpha$ column of Table 3.5, multiplying each number by the corresponding r. The first value of r that produces a $r \times k_{r;\,1-\alpha}$ greater than our target value is 1 higher than the desired pass criteria.

EXAMPLE 3.11 CHOOSING PASS/FAIL CRITERIA

What is the maximum number of failures we can allow in order to be 80% confident of a failure rate no higher than 5%/K, if 50 units are to be tested for 2000 hr?

SOLUTION

The basic equation is $[r/(50 \times 2000)] \times k_{r;\,1-\alpha} = .00005$. From this we obtain $r \times k_{r;\,1-\alpha} = 5$. Looking down the 80% column of Table 3.5 we calculate $1 \times 2.99 = 2.99$, $2 \times 2.14 = 4.28$, $3 \times 1.84 = 5.52$. The last product, for $r = 3$, is greater than 5. Therefore, we subtract 1 from 3 and come up with a pass criteria of up to 2 failures on the test.

Case III. What Are Minimum Test Sample Sizes We Can Use

As before, we have to specify a failure rate objective and a confidence level and a test duration. For a minimum sample size, we anticipate the best possible outcome, namely, 0 failures. This means we are prepared to state the product has not demonstrated the specified failure rate at the required confidence level if we see even 1 fail.

Minimum sample sizes are derived by setting $k_{0;1-\alpha}/nT = \lambda_{\text{obj}}$, where $k_{0;1-\alpha}$ is the zero failures factor from Table 3.7. Solving for n gives $n = k_{0;1-\alpha}/(\lambda_{\text{obj}} \times T)$. An equivalent formula that works for any α level without needing tables is

$$n = \frac{-\ln \alpha}{\lambda_{\text{obj}} \times T}$$

EXAMPLE 3.12 MINIMUM SAMPLE SIZES

We want a minimum sample size that will allow us to verify a 40,000-hr MTTF with 90% confidence, given the test can last 8000 hr.

SOLUTION

The $k_{0;.9}$ factor is 2.3026 (from Table 3.7). Therefore, $n = (40000 \times 2.3026)/8000 = 12$ (after rounding up).

Variation: How Do We Determine Minimum Testing Times?

The number of test units, as well as the failure rate objective and the confidence level, are fixed in advance. The choice of T then becomes

$$T = k_{0;1-\alpha}/(n \times \lambda_{obj}) = -\ln \alpha/(n \times \lambda_{obj}).$$

As before, if 1 fail occurs when the test is run, the failure rate objective will not be confirmed at the desired confidence level.

EXAMPLE 3.13 MINIMUM TEST TIMES

The failure rate objective is the very low number 10 PPM/K. We want to confirm this at an 80% confidence level. The component is an inexpensive resistor, and we plan to test 10,000 of them. How long should the test run?

SOLUTION

By substitution, $T = 1.6094/(10{,}000 \times 10 \times 10^{-9}) = 16{,}094$ hr.

This period is about two years of continuous testing, and might be much too long to be practical, even though it is a "minimum" test time. This illustrates the difficulties inherent in verifying very high levels of reliability. Since the trend is toward more and more failure rate objectives in the PPM/K range, reliability analysts will face this problem with increasing frequency. The better we make our components, the harder it becomes to assess their actual performance.

One way out of this quandary is to test at high levels of stress, accelerating failure times as compared to what would happen at actual use conditions. We already saw this done (but not mathematically explained) with the light bulb equivalent month of failure data (Example 3.3). A full discussion of acceleration modeling is in Chapter 7. At this point we just note that if the test in Example 3.13 could be carried out in a test condition that accelerates failure times by a factor of 10 (i.e., a 10 × acceleration factor), then only 1604 test hours would be required, or under 10 weeks.

Using a minimum sampling plan will generally save around 40% in sample size or test duration, as compared to allowing just 1 fail. On the other hand, deciding a product will not meet its objective based on a single fail leaves no margin for the one odd defective unit that might have slipped into the sample. It may turn out better to allow for a few failures, and pay the extra sample size price, rather than have a test result that many will not accept as valid. Use of minimum sample sizes makes sense if we are very confident the product is much better than the objective, or we think it is so much worse that we will see many failures, even with the minimum sample.

SUMMARY

The exponential life distribution is defined by $F(t) = 1 - e^{-\lambda t}$. It is characterized by being the only distribution to have a constant failure rate. This constant failure rate has the same value as the one unknown parameter λ, which is also the reciprocal of the mean time to fail (MTTF).

Another important characterization property of the exponential is its "lack of memory." When failures seem to occur randomly at a steady rate, with no significant wearout or degradation mechanisms, then the exponential will be a good model. It is appropriate for the long flat portion of the widespread "bathtub" curve for failure rates. It is also useful when we want to verify

Table 3.8. Exponential Distribution Formulas and Properties.

NAME	VALUE OR DEFINITION
CDF $F(t)$	$1 - e^{-\lambda t}$
Reliability $R(t)$	$e^{-\lambda t}$
PDF $f(t)$	$\lambda e^{-\lambda t}$
Failure rate $h(t)$ or renewal rate	λ
AFR (t_1, t_2)	λ
Mean $E(t)$ or MTTF (also MTBF)	$1/\lambda$
Variance $V(t)$	$1/\lambda^2$
Median	$\frac{.693}{\lambda}$
Mode	0
Lack of memory property	The probability a component fails in the next t_2 hours, after operating successfully for t_1 hours, is the same as the probability of a new component failing in its first t_2 hr.
Closure property	A system of n independent components each exponentially distributed with parameters $\lambda_1, \lambda_2, \ldots, \lambda_n$, respectively, has an exponential distribution for the time to first failure, with parameter $\lambda_s = \lambda_1 + \lambda_2 + \ldots + \lambda_n$.

an average failure rate over an interval, and the time pattern of fails over the interval is of little concern.

The key formulas and properties of the exponential distribution are summarized in Table 3.8, for easy reference.

The best estimate of λ from censored Type I or II data is the number of fails divided by the total hours on test of all units. For readout data, the denominator may not be known exactly. Even in this case, however, the total unit test hours can often be estimated with little loss of precision.

When λ has been estimated from censored Type I or II data, there are factors which only depend on the number of fails that can be used to obtain upper and lower bounds on λ. These are given in Tables 3.5 and 3.6. Table 3.7 has formulas for upper bounds when there are no fails on the test.

The factors in Tables 3.5 and 3.7 can also be very useful in the planning stages of an experiment. Sample sizes, or test durations, can be calculated once the failure rate objective and a confidence level are specified. The zero fail formulas yield minimum sample sizes. Even when the unit to be tested does not have an exponential distribution, the numbers obtained from these tables can be used for rough planning.

When data are available, a histogram plotted on the same graph as the estimated PDF $f(t)$ (properly scaled up) gives a good visual check on the adequacy of using an exponential model. In addition, the chi-square goodness of fit test provides an analytic check.

4
The Weibull Distribution

In the last chapter, we saw simple, yet powerful methods for analyzing exponential data and planning life test experiments. Questions about sample size selection, test duration, and confidence bounds could all be answered using a few tables. However, these methods apply only under the constant failure rate assumption or the equivalent "lack of memory" property. As long as this assumption is nearly valid over the range of failure times we are concerned with, we can use the methods given. But what do we do when the failure rate is clearly decreasing (typical of early failure mechanisms), or increasing (typical of later life wearout mechanisms)?

This problem was tackled by Waloddi Weibull (1951). He derived the generalization of the exponential distribution that now bears his name. Since that time, the Weibull distribution has proved to be a successful model for many product failure mechanisms, because it is a flexible distribution with a wide variety of possible failure rate curve shapes. In addition, the Weibull distribution also has a derivation as a so-called "extreme value" distribution which suggests its theoretical applicability when failure is due to a "weakest link" of many possible failure points.

First we will derive the Weibull as an extension of the exponential. Then we will discuss the extreme value theory as an added bonus: not only does the Weibull appear to "work" in many practical applications, there is also an explanation to tell us why it works and in what areas it is likely to be most successful.

EMPIRICAL DERIVATION OF THE WEIBULL DISTRIBUTION

The goal is to find a CDF that has a wide variety of failure rate shapes, with the constant $h(t) = \lambda$ as just one possibility. Allowing any polynomial form of the type $h(t) = at^b$ for a failure rate function would accomplish this.

In order to derive $F(t)$, it is easier to start with the cumulative failure rate function $H(t)$. Setting

$$H(t) = (\lambda t)^m$$

gives us the exponential constant failure rate when $m = 1$, and a polynomial failure rate for other values of m. This follows since

$$h(t) = dH(t)/dt = m\lambda(\lambda t)^{m-1}.$$

Now we use the basic identity relating $F(t)$ and $H(t)$

$$F(t) = 1 - e^{-H(t)} = 1 - e^{-(\lambda t)^m}.$$

We obtain the form for the Weibull CDF we shall use by making a substitution of $c = 1/\lambda$ in the above equation.

$$F(t) = 1 - e^{-(t/c)^m}.$$

The parameter c is called the characteristic life. The parameter m is known as the shape parameter. Both c and m must be greater than zero and the distribution is a life distribution defined only for positive times t.

The PDF and failure rate $h(t)$ and AFR for the Weibull are given by

$$f(t) = \frac{m}{t}\left(\frac{t}{c}\right)^m e^{-(t/c)^m}$$

$$h(t) = \frac{m}{c}\left(\frac{t}{c}\right)^{m-1} = \frac{m}{t}\left(\frac{t}{c}\right)^m$$

$$\text{AFR}(t_1, t_2) = \frac{(t_2/c)^m - (t_1/c)^m}{t_2 - t_1}$$

$$\text{AFR}(T) = \frac{1}{c}\left(\frac{T}{c}\right)^{m-1}$$

There is, unfortunately, no consistent convention used throughout the literature when naming the Weibull parameters. Often the shape parameter is known as β. The characteristic life parameter c may be designated by α or η and is sometimes called the scale parameter. There is also an alternative form of the Weibull often encountered where the scale parameter $\theta = c^m$ is used resulting in

$$F(t) = 1 - e^{-t^m/\theta}.$$

Because of this confusion of terminology and parameterization, the reader should be careful about the definitions used when reading literature about the Weibull.

EXAMPLE 4.1 WEIBULL PROPERTIES

A population of capacitors are known to fail according to a Weibull distribution with characteristic life $c = 20{,}000$ power-on hours. Evaluate the probability a new capacitor will fail by 100, 1000, 20,000, and 30,000 hr, for the cases where the shape parameter m is .5 or 1.0 or 2.0. Also calculate the failure rates at these times for these three shape parameters.

SOLUTION

Table 4.1 gives the various values requested. Several things are worth noting about the results. In particular, observe how when the shape value is .5, the CDF at 100 hr is much higher than when $m = 1$ or $m = 2$. Also, the failure rate values for $m = .5$ are highest at the early time, and decrease with each later time. Exactly the opposite is true for the failure rate values when $m = 2$.

Table 4.1 also shows, as the Weibull derivation given in this section indicates, that the failure rate for $m = 1$ is a constant. This implies the Weibull reduces to an exponential with $\lambda = .00005 = 1/c$.

Finally, note that the percent fail values at the characteristic lifetime value of 20000 hr were uniformly 63.2, for all the choices of m.

All these observations based on Table 4.1 illustrate general points about the Weibull distribution which will be discussed in the next section.

The Weibull CDF equation has four quantities that may be known, assumed, or estimated from data. These are: the cumulative fraction failed

Table 4.1. Solution to Example 4.1.

	CDF IN %			FAILURE RATE IN %/K		
TIME	$m = .5$	$m = 1.0$	$m = 2.0$	$m = .5$	$m = 1.0$	$m = 2.0$
100	7.0	.5	.002	35.4	5.0	.05
1,000	20.0	4.9	.2	11.2	5.0	.5
20,000	63.2	63.2	63.2	2.5	5.0	10.0
30,000	70.6	77.7	89.5	2.0	5.0	15.0

$F(t)$; the time t; the shape parameter m; and the characteristic life parameter c. If any three of these are known the fourth can be calculated by one of the equations below.

$$F(t) = 1 - e^{-(t/c)^m}$$

$$m = \frac{\ln[-\ln(1 - F)]}{\ln(t/c)}$$

$$t = c[-\ln(1 - F)]^{1/m}$$

$$c = \frac{t}{[-\ln(1 - F)]^{1/m}}.$$

PROPERTIES OF THE WEIBULL DISTRIBUTION

The strength of the Weibull lies in its flexible shape as a model for many different kinds of data. The shape parameter m plays the major role in determining how the Weibull will look. For $0 < m < 1$, the PDF approaches infinity as time approaches zero, and is always decreasing rapidly toward zero as time increases. The failure rate behaves the same way, making this type of Weibull an ideal model for an early failure mechanism typical of the front end of the "bathtub" curve.

When $m = 1$, the Weibull reduces to a standard exponential with constant failure rate $\lambda = 1/c$.

For $m > 1$, the PDF starts at zero and increases to a peak at $c[1 - (1/m)]^{1/m}$. From then on it decreases toward zero as time increases. The shape is skewed to the right. The failure rate also starts at zero but then increases monotonically throughout life. The rate of increase depends on the size of m. For example, if m is 2, the failure rate increases linearly (and the distribution is also known as the Rayleigh distribution). When m is 3, the failure rate has a quadratic rate of increase, and so on. This type of Weibull is a useful model for wearout failure mechanisms typical of the back end of the "bathtub" curve.

Figure 4.1 shows several examples of Weibull PDFs, and Figure 4.2 shows a variety of Weibull failure rate (hazard) curves. Table 4.2 summarizes the way the Weibull varies according to the value of its shape parameter m.

The second parameter of the Weibull, c, is a scale parameter that fixes one point of the CDF; the 63.2 percentile or characteristic life point. This follows by substituting c for time in the CDF and obtaining $F(c) = 1 - e^{-(c/c)^m} = 1 - e^{-1} = .632$. In other words, 63.2% of the population fails by the characteristic life point, independent of the value of the shape parameter m.

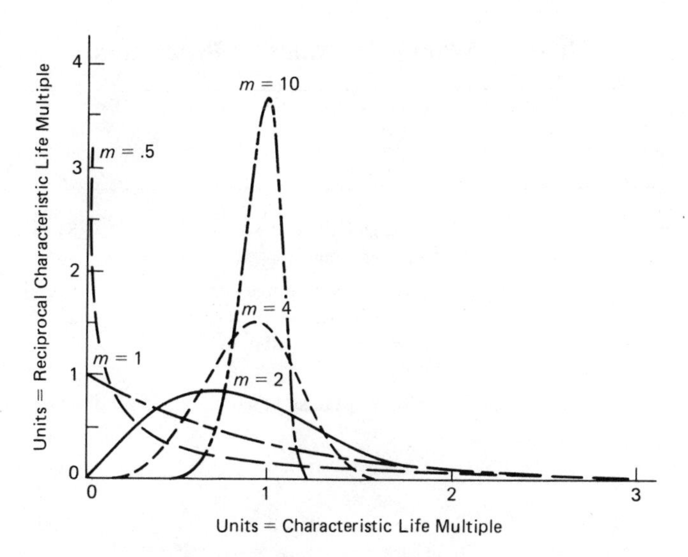

Figure 4.1. Weibull Probability Density.

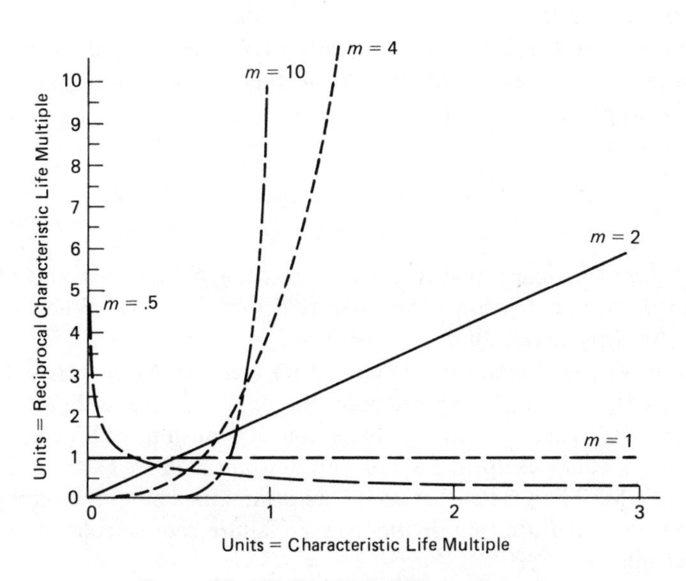

Figure 4.2. Weibull Failure Rate (Hazard).

Table 4.2. Weibull Distribution Properties.

SHAPE PARAMETER, m	PDF	FAILURE RATE, $h(t)$
$0 < m < 1$	Exponentially decreasing from infinity	Same
$m = 1$	Exponentially decreasing from $1/c$ (exponential distribution)	Constant at $1/c$
$m > 1$	Rises to peak and then decreases	Increasing
$m = 2$	Rayleigh distribution	Linearly increasing
$3 \le m \le 4$	Has "normal" bell shape appearance	Rapidly increasing
$m > 10$	Has shape very similar to Type I extreme value distribution	Very rapidly increasing

The median, or T_{50} point for the Weibull is found by letting $F(T_{50}) = .5$ to yield $T_{50} = c(\ln 2)^{1/m}$.

In order to give the mean and the variance of the Weibull distribution, it is necessary to introduce a mathematical function known as the gamma function. This function is defined by

$$\Gamma(v) = \int_0^\infty y^{v-1} e^{-y}\, dy$$

and is tabulated in many places, for example, see Abramovitz and Stegun (1964). In particular, when v is an integer, $\Gamma(v) = (v - 1)!$. When $v = .5$, $\Gamma(.5) = \sqrt{\pi}$. In general, for $v > 0$, $\Gamma(v) = (v - 1)\,\Gamma(v - 1)$.

The mean of the Weibull is $c\Gamma(1 + 1/m)$ and the variance is $c^2\Gamma(1 + 2/m) - [c\Gamma(1 + 1/m)]^2$. For example, for $m = .5$, 1, and 2, the means are $2c$, c, and $\sqrt{\pi}/2c$, respectively. Note that this mean or MTTF no longer has any direct relationship to the failure rate (unless $m = 1$). So while the Weibull has a MTTF, it is not that useful or meaningful a number as compared to a graph of the failure rate or the average failure rate calculated over an interval of interest.

The Weibull also has a closure or reproductive property, similar to the exponential. If a system is composed of n parts, each having an independent Weibull distribution with the same shape parameter but not necessarily the same characteristic lives, then the time to the first system fail also follows a Weibull distribution. If the characteristic life parameters are $c_1, c_2, \ldots, c_n$, and the shape parameter is m, then the first system fail distribution also has shape parameter m. The system characteristic life is given by

$$c_s = \left(\sum_{i=1}^{n} \frac{1}{c_i^m} \right)^{-\frac{1}{m}}.$$

EXAMPLE 4.2 WEIBULL CLOSURE PROPERTY

A car manufacturer uses five different hoses as a part of the engine cooling system for one of its models. The hose manufacturer specifies that each hose has a lifetime modeled adequately by a Weibull distribution with shape parameter 1.8. The five hoses have characteristic lifes, in months of average car use, of 95, 110, 130, 130, and 150. What is the life distribution for time to

Table 4.3. Weibull Formulas.

NAME	VALUE OR DEFINITION
CDF $F(t)$	$1 - e^{-(t/c)^m}$
Reliability $R(t)$	$e^{-(t/c)^m}$
PDF $f(t)$	$(m/t)(t/c)^m e^{-(t/c)^m}$
Characteristic life c	$F(c) = 63.2$
Shape parameter m	Determines shape of Weibull
Failure rate $h(t)$	$m/c\,(t/c)^{m-1}$
AFR (t_1, t_2)	$\dfrac{[(t_2/c)^m - (t_1/c)^m]}{t_2 - t_1}$
Mean $E(t)$	$c\Gamma(1 + 1/m)$
Variance $V(t)$	$c^2\Gamma(1 + 2/m) - [c\Gamma(1 + 1/m)]^2$
Median T_{50}	$c(\ln 2)^{1/m}$
Mode	$c(1 - 1/m)^{1/m}$
System characteristic life c_s when components are independent Weibull with parameters (m, c_i)	$c_s = \left(\sum_{i=1}^{n} \frac{1}{c_i^m} \right)^{-\frac{1}{m}}$

first car hose failure? What is the MTTF? What is the median or T_{50}? How likely is it no hose fails in the first year of car life? By 4 years?

SOLUTION

Applying the closure property relationship, we have that the hose system (first) failure distribution is a Weibull with shape parameter $m = 1.8$ and characteristic life

$$c_s = \left(\frac{1}{95^{1.8}} + \frac{1}{110^{1.8}} + \frac{1}{130^{1.8}} + \frac{1}{130^{1.8}} + \frac{1}{150^{1.8}}\right)^{-1/1.8} = 48.6.$$

The MTTF is $c\Gamma(1 + 1/1.8) = 43.2$ months. The T_{50} is $c(\ln 2)^{1/m} = 39.6$ months. The reliability at 12 months, or $[1 - F(12)]$, is 92%. The reliability at 48 months is approximately (100 − the characteristic life percent), or 37%.

The parameter definitions and key Weibull formulas given in this section are summarized in Table 4.3.

EXTREME VALUE DISTRIBUTION RELATIONSHIP

In general, if we are interested in the minimum of a large number of similar independent random variables, as for example, the first time of failure reached by many competing similar defect sites located within a material, then the resulting life distribution can converge to only one of three types. These types, known as smallest extreme value distributions, depend on whether we define the random variables on the entire x axis, or just the positive or negative halves of the axis.

The extreme value distributions were categorized by Gumbel in 1958 and are

$$F(x) = 1 - e^{-e^{x}} \qquad -\infty < x < \infty \qquad \text{(Gumbel Type I)}$$

$$F(x) = 1 - e^{-(-x)^{-a}} \qquad x \leq 0$$

$$F(x) = 1 - e^{-x^{a}} \qquad x \geq 0 \qquad \text{(Weibull).}$$

In the above formulas, the random variables have been suitably shifted by any location parameters and divided by scale parameters so as to appear "normalized." Of the three possible types, the only one that is a life distribution (i.e., defined only for nonnegative values) turns out to be the Weibull. Since failure can often be modeled as a weakest link of many competing failure processes, the wide applicability of the Weibull is not surprising.

The justification for using the Weibull based on extreme value theory is an important and useful one. It can also be abused and lead to vigorously defended misapplications of the Weibull. For this reason, the strengths and weaknesses of this derivation should be carefully noted. If there are *many identical and independent competing processes* leading to failure, and the *first* to reach a critical stage determines the failure time, then—*provided "many" is large enough*—we can derive a Weibull distribution. In a typical modeling application, we might suspect or hope all the underlined assumptions apply, but we are not likely to be certain.

How do we pick a life distribution model from a practical viewpoint? The approach we recommend is as follows: Use a life distribution model primarily because it works, that is, *fits the data well and leads to reasonable projections when extrapolating beyond the range of the data.* Look for a new model when the one previously used no longer "works." Select models by researching which models have been used in the literature successfully for similar failure mechanisms, or by using theoretical arguments applied to the physical models of failure. For example, extreme value theory, the multiplicative degradation model theory of the next chapter on the lognormal distribution, or the "lack of memory" exponential model of the preceding chapter, are properties to consider.

Another situation where theoretical derivations prove very useful is when several distributions all seem to work well with the available data, but give significantly different results when projected into critical tail regions that are beyond the range of the data. In this case, pick the model that has a theoretical derivation most closely matching the cause of failure.

There is another very interesting mathematical relationship between the Weibull distribution and the Type I extreme value distribution. It turns out that the natural logarithms of a population of Weibull failure times form a population following the Type I distribution. In other words, the natural logarithm of a Weibull random variable is a random variable that has the Type I extreme Value CDF. This relationship, as we shall see in the next chapter, is exactly the same as exists between the lognormal life distribution and the normal distribution. So, if the Weibull had not been named in honor of its chief popularizer, it probably would have been called the log-extreme value distribution.

The exact statement of this relationship is as follows: let t_f be a Weibull random variable with CDF

$$F(t) = 1 - e^{-(t/c)^m}.$$

Then the random variable $X = \ln t_f$ has the Type I extreme value CDF given by

$$1 - e^{-e^{(x-a)/b}}$$

with $a = \ln c$ and $b = 1/m$.

A reliability analyst who has computer programs or graph paper designed to handle Type I extreme value data can also analyze Weibull data after first transforming it into extreme value data (by taking natural logarithms). Later on the scale parameter b can be used to estimate the Weibull shape via $m = 1/b$. The location parameter a is transformed into the Weibull characteristic life via $c = e^a$.

AREAS OF APPLICATION

After the introduction of the Weibull distribution, its use spread across a wide variety of applications, running the range from vacuum tubes and capacitors to ball bearings and relays and material strengths. The primary justification for its use has always been its flexible ability to match a wide range of phenomena. There are few, if any, observed failure rates that cannot be accurately described over a significant range of time by a polynomial or Weibull hazard function.

Some particular applications, such as modeling capacitor dielectric breakdown, fit nicely into the "worst link," or first of many flaws to produce a failure, extreme value theory. Dielectric materials contain many flaws, all "competing" to be the eventual catastrophic failure site. In many cases, the failures occur mostly early in life and a Weibull with a shape parameter less than 1 works best.

On the other hand, there is less reason to expect a Weibull to apply when failure is due to a chemical reaction or a degradation process such as corrosion or migration or diffusion (although even here the many competing sites argument might still possibly apply). It is in precisely such applications, typical of many semiconductor failure mechanisms, that the lognormal distribution (Chapter 5) has replaced the Weibull as the most popular distribution.

One particular form of Weibull deserves special mention. When $m = 2$, as noted in Table 4.2, the distribution is called the Rayleigh. The failure rate increases linearly with $h(t) = (m/c^2)t$ and the CDF is given by

$$F(t) = 1 - e^{-(t/c)^2}.$$

There is an interesting measurement error problem which also leads to this same CDF.

Assume you are measuring, or trying to locate, a particular point on the plane. A reasonable model that is often used measures independent x and y coordinates. Each measurement has a random amount of error, modeled as usual by the normal distribution (Chapter 5). Assume each error distribution has zero mean and the same standard deviation, σ. If the error in the x direction is the random variable X, and the y direction error is Y, then the total radial error (or distance from the correction location) is $R = \sqrt{X^2 + Y^2}$. Using standard calculus methods, the CDF of R can be derived. It turns out to be

$$F(r) = 1 - e^{-(r^2/2\sigma^2)}$$

which is the Rayleigh distribution with $c = \sqrt{2}\sigma$.

EXAMPLE 4.3 RAYLEIGH RADIAL ERROR

An appliance manufacturer wants to purchase a robot arm to automate a particular assembly operation. The quality organization has been asked to evaluate the reliability of an arm under consideration. A key point in the evaluation is whether the arm can repeatedly go to specified points in its operating range, within a tolerable margin of positioning error. The literature on the arm says that it will repeatedly arrive at programmed points with an accuracy of plus or minus .3 cm in either the x coordinate or the y coordinate direction.

Test have determined that the operation will succeed as long as the arm arrives no further than .4 cm from the designated point. If an error rate of less than 1 in 1000 is required, will the arm under consideration be adequate?

SOLUTION

This example is instructive, not only of the Rayleigh distribution, but also of the kind of detective work a statistician or reliability analyst must often carry out. What does an accuracy of plus or minus .3 cm really mean? Often a phone call to the company will not produce an immediately satisfactory answer. In the meantime, an analyst can make an evaluation based on making standard assumptions. These will involve a knowledge of the normal distribution, discussed in the next chapter.

Assume that plus or minus .3 cm refers to plus or minus three standard deviations (or sigmas) of the typical normal error distribution. Sigma is then .1 cm. If we also assume that the placement errors in each coordinate are independent with an average value of zero, the point the robot arm arrives at will have a random distance from the objective point with a CDF given by the Rayleigh distribution

$$F(r) = 1 - e^{-r^2/2(.1)^2} = 1 - e^{-50r^2}.$$

Substituting .4 for r yields $F(.4) = .9997$. This means the arm will be more than .4 cm off only about 3 times every 10,000 operations, which meets the less than 1 in 1000 objective.

WEIBULL PARAMETER ESTIMATION

Two analytic methods for estimating m and c from data (either complete or censored or grouped samples) will be described in this section. The most recommended procedure is called the method of maximum likelihood. It is a standard, well-known technique, described fully in most statistics textbooks (for example, see Wilks, 1962). Its use for censored reliability data is described in detail in Nelson (1982) and also in Mann et al. (1974). Unfortunately, its implementation for censored or grouped data is generally not practical unless appropriate computer programs are available to do the calculations. These programs may be purchased by the reader as a part of statistical analysis packages such as Censor (described by Meeker and Duke, 1981) or Statpac (developed by W. Nelson and described by Strauss, 1981). The reader with programming experience can write his own procedures from the equations given in Nelson (1982). The second estimation method is easier to implement. It is an analytic version of the graphical estimation techniques described in Chapter 6. Only the common least-squares or regression programs found in many hand-held calculators are needed.

At this point it is natural to inquire why a complicated method requiring fairly rare computer programs is recommended when simple, intuitive graphical techniques, combined with a curve-fitting routine that eliminates subjective judgment, can be used. Not only that, but the graphical approach (as we shall see in the next section) offers an immediate visual test of whether the Weibull distribution fits the data or not (based on whether the data points line up in an approximate line on special Weibull graph paper). The reason for the recommendation has to do with the concept of accuracy, in a statistical sense.

In a real life problem, when we estimate parameters from data we can come up with many often widely differing results, depending on the estimation method we use. Important business decisions may depend on which estimate we choose. Obviously, it is important to have objective criteria that tell us which method is best in a particular situation. Accuracy, in a statistical sense, starts with a definition of desirable properties an estimation method may have and continues with an investigation into which of these "good" properties the methods available actually have. Note that we are talking about the method of estimation—not the estimates that can be derived from the use of these methods on a specific set of data. Due to the random nature of sample data, an almost arbitrary guess based on looking at a few of the data points might, on any given day, turn out to be closer than the estimate obtained from the most highly recommended computer program. Statistical theory describes how well various methods compare to each other in the long run—over many, many applications. We have to accept the logic that it makes sense to use the most reliable long-run method on any single set of data we want to analyze.

The most desirable attributes defined for estimation methods are the following:

1. Lack of bias: The expected value of the estimate equals the true parameter (or, on average, you're centered "on target")
2. Minimum variance: A minimum variance estimator has less variability on the average than any other estimator and, if also unbiased, is likely to be closer to the true value than another estimator.
3. Sufficiency: The estimate makes use of all the statistical information available in the data.
4. Consistency: The estimate tends to get closer to true value with larger size samples ("infinite" samples yield perfect estimates).

In addition, we want estimates to have a known distribution that can be utilized for forming confidence intervals and carrying out tests of hypotheses.

In general, no known method provide all of the attributes mentioned. Indeed, it may be difficult to find any method with a lack of bias or minimum variance when dealing with life distributions and censored data. The maximum likelihood method can be shown to possess all the above properties as sample sizes (or numbers of failures) become large enough. This property, called asymptotic behavior, assures us that, for reasonable amounts of data, no other estimation technique is "better." Asymptotic theory does not tell us "how large" is "large" but practical experience and simulation experiments indicate that more than 20 failures is "large" and, typically, if there are over 10 failures the maximum likelihood estimates (MLEs) are accurate. For smaller amounts of data, the unbiased minimum variance property can not be claimed, but better techniques are hard to come by.

A loose but useful description of the MLE technique is as follows: the "probability" of the sample is written by multiplying the density function evaluated at each data point. This product, containing the data points and the unknown parameters, is called the likelihood function. By finding parameter values that maximize this expression we make the set of data observed "more likely." In other words, we choose parameter values that are most consistent with our data by maximizing the likelihood of the sample.

The MLE technique is therefore equivalent to maximizing an equation of several variables. In general, the standard calculus approach of taking partial derivatives with respect to each of the unknown parameters and setting them equal to zero, will yield equations that have the MLEs as solutions. In most cases, by first taking natural logarithms of the likelihood equation, and then taking partial derivatives to solve for a maximum, the calculations are simplified. The same parameter values that maximize the log likelihood will, of course, maximize the likelihood. However, for censored or grouped data, these equations are nonlinear and complicated to set up and solve. Consequently, appropriate computer programs are needed.

When the life distribution is the exponential, the MLE equations are easy to derive and solve, even for censored data. This procedure is shown in the next example, which illustrates how the MLE method works.

EXAMPLE 4.4 MLE FOR THE EXPONENTIAL

Show that the estimate of the exponential parameter λ given in Chapter 3 (i.e., the number of failures divided by the total unit test hours) is the MLE estimate for complete or censored Type I (time censored) or censored Type II (rth fail censored) data.

SOLUTION

The likelihood equation is given by

$$\mathrm{LIK} = k\left(\prod_{i=1}^{r} f(t_i)\right)(1 - F(T))^{n-r}$$

or

$$\mathrm{LIK} = k\lambda^r \left(e - \lambda \sum_{i=1}^{r} t_i\right)(e^{-\lambda T})^{n-r},$$

where k is a constant independent of λ and not important for the maximizing problem. The last term in LIK is the probability of $n - r$ sample units surviving past the time T. If T is fixed in advance, we have Type I censoring. If T is the time of the rth fail, we have Type II censoring. If $r = n$, the sample is complete or uncensored.

If, we let L denote the log likelihood (without any constant term) then

$$L = r \ln \lambda - \lambda \sum_{i=r}^{r} t_i + (n - r)(-\lambda T).$$

To find the value of λ which maximizes L we take the derivative with respect to λ and set it equal to 0.

$$\frac{dL}{d\lambda} = \frac{r}{\lambda} - \sum_{i=1}^{r} t_i - (n - r)T = 0.$$

Solving for λ, we have the estimate given in Chapter 3.

$$\hat{\lambda} = \frac{r}{\sum_{i=1}^{r} t_i + (n - r)T.}$$

If the data are readout or grouped, the MLE, even in the simple exponential case, is not easy to obtain. The likelihood equation is LIK $= k\ F(t_1)^{r_1} \left\{\prod_{i=2}^{m} [F(t_i) - F(t_{i-1})]^{r_i}\right\} [1 - F(t_m)]^{n-r}$, where $t_1, t_2, \ldots, t_m$ are the readout times and $r_1, r_2, \ldots, r_m$ are the failures first observed at those times and F is the CDF for the assumed life distribution and $r = \sum_{i=1}^{m} r_i$ is the total number of failures out of n on test. The partial derivative equations that result in the Weibull case may be found in Chace (1976). For multicensored data, with l_i units removed from test at time L_i, add a factor term to LIK of the form $[1 - F(L_i)]^{l_i}$. This term reflects the fact that we know only that those units have survived to time L_i.

A very useful part of the large sample theory for MLE estimates is that they have an asymptotic normal distribution. The mean is the true parameter value and the standard deviation can be estimated from equations based on partial derivatives of the log likelihood equations. This theory will not be given here: it is described in detail by Nelson (1982), and the calculations should be part of any program that obtains MLEs from reliability data.

One more use of MLEs deserves mention. As stated in Chapter 3, when doing a goodness of fit test to check a distribution assumption, the type of estimates to use in place of unknown parameters are MLEs.

The second estimation method for Weibull parameters is based on a procedure called "linear rectification." The idea is to put the Weibull CDF equation into a form that, with the proper substitution of variables, is linear. The equations are

$$F(t) = 1 - e^{-(t/c)^m}$$

$$\ln[1 - F(t)] = -(t/c)^m$$

$$\ln\{-\ln[1 - F(t)]\} = m \ln t - m \ln c$$

$$Y = mX + b.$$

In the final linear form, Y is the estimated value of $\ln\{-\ln[1 - F(t)]\}$ at time t, and X is the natural log of t. To estimate Y, we have to estimate $F(t)$. The estimate to use, as described in Chapter 6, depends on whether exact times of failure are available [in that case, use $F(t_i) = (i - .3)/(n + .4)$ for the ith failure time], or whether the data are readout; use $F(t_i) =$ (total number of failures up to time t_i)$/n$.

When the calculations are completed, there is an (X, Y) pair for each data point or readout time. A least-squares fit, or regression of Y on X, yields estimates for the slope m and the intercept $b = -m \ln c$. Estimate c by $c = e^{-b/m}$. Programs or calculators that will accept the (X, Y) pairs as inputs, and give least-squares estimates of m and b as outputs, are common. (An elementary discussion of using the least-squares method to fit a line to data points, and linear rectification, is given in Chapter 6.)

Weibull graph paper, discussed in Chapter 6, has scales adjusted in the same way as the transformations that obtained (X, Y) from the CDF estimate and the time of fail. Therefore, a plot of the CDF estimates versus time on this specially constructed paper will yield an approximate straight line with slope m and intercept $m \ln c$, provided the data follow a Weibull model. The computer least-squares procedure based on linear rectification is an objective way to put a line on Weibull graph paper which minimizes the squared deviations in the cumulative percent failure scale direction.

In many statistical applications, the estimates obtained via least-squares or regression methods can be shown to have very desirable properties similar to those described for MLEs. However, there are several key assumptions about the (X, Y) points which must be made in order for the least-squares "optimality" properties to hold. Basically, the random errors in Y at the X points must be uncorrelated and have zero average value and the same variance. All three of these assumptions are known not to hold in the application of least squares described above. This method of estimating Weibull parameters gives convenient analytic estimates, which will be good for large amounts of data (consistent property). Little else can be said for them in terms of desirable properties. Also, any confidence bounds on m or b given in the output of a regression program are not valid for this application.

EXAMPLE 4.5 WEIBULL PARAMETER ESTIMATION

Capacitors, believed to have operating lifetimes modeled adequately by a Weibull distribution, were tested at high stress to obtain failure data. Fifty units were run and 25 failed by the end of the test. First assume the failure times were continuously monitored and the times of fail were, in hours to the nearest tenth, .7, 52.7, 129.4, 187.8, 264.4, 272.8, 304.2, 305.1, 309.8, 310.5, 404.8, 434.6, 434.9, 479.2, 525.3, 620.3, 782.8, 1122, 1200.8, 1224.1, 1322.7, 1945, 2419.5, 2894.5, 2920.1. Assume the test ended at the last fail and estimate the Weibull parameters. Next, assume only readout data were taken. The readouts were made at 24, 168, 200, 400, 600, 1000, 1500, 2000, 2500, and 3000 hr. The new failures observed corresponding to these readout times were 1, 2, 1, 6, 5, 2, 4, 1, 1, and 2. Again, estimate m and c.

SOLUTION

An MLE program yields $\hat{m} = .62$ and $\hat{c} = 5078$ for the exact fail time data. Using the regression method, the Y and X values shown below are inputted into a least-squares line-fitting program yielding $\hat{m} = .55$ and $\hat{b} = -4.8277$. This gives $\hat{c} = e^{4.8277/.55} = 6488$.

Y	X
−4.27	−.36
−3.37	3.96
−2.90	4.86
−2.57	5.24
−2.32	5.58
−2.12	5.61
−1.95	5.72
−1.80	5.72
−1.66	5.74
−1.54	5.74
−1.43	6.00
−1.33	6.07
−1.24	6.08
−1.15	6.17
−1.06	6.26
−.99	6.43
−.91	6.66
−.84	7.02
−.77	7.09
−.70	7.11
−.64	7.19
−.57	7.57
−.51	7.79
−.45	7.97
−.40	7.98

For the readout data, MLE estimates of m and c are $\hat{m} = .64$ and $\hat{c} = 5025$. The least-squares method uses the X and Y values below to obtain $\hat{m} = .78$ and $\hat{c} = 3759$.

Y	X
−3.90	3.18
−2.78	5.12
−2.48	5.30
−1.50	5.99
−1.03	6.40
−.88	6.91
−.61	7.31
−.55	7.60
−.48	7.82
−.37	8.01

Since the data in Example 4.5 were simulated we can compare the various estimates obtained to the true Weibull parameter values: $m = .6$ and $c = 4000$. The closest estimates for the critical m parameter are given by the MLE method. The least-squares method with interval data happens, in this example, to come closest to the correct c value.

CDF values calculated using either of the above estimation methods will seldom differ significantly for times within the range where experimental failures occurred. However, we are often concerned with extrapolating back to very early times in the front tail of the life distribution—percentiles much smaller than experimental sample sizes allow us to observe directly. Here, small changes in estimated parameter values, especially in the shape parameter, can make orders of magnitude difference in the CDF estimate. Hence, it is important to use the best technique available for final parameter estimates.

Chapter 6 will describe techniques for plotting Weibull data and obtaining quick parameter estimates. Even though these estimates are not recommended for use in critical applications, the value obtained from looking at your data on the appropriate graph paper can not be overemphasized. Sometimes a strange pattern of points on graph paper may lead an analyst to ask questions that lead to a valuable insight that would have been lost had the entire analysis been done by computer programs. Even if the plot only serves to confirm the model chosen, it is useful for presentation and validation purposes.

SUMMARY

The Weibull distribution, with CDF $F(t) = 1 - e^{-(t/c)^m}$, is a flexible, convenient life distribution model. It has a family of polynomial shaped failure rate functions, depending on the value of the shape parameter m, including decreasing, constant, and increasing failure rates. Many components and systems have reliability properties that are successfully modeled by the Weibull.

The Weibull also has a theoretical derivation as an extreme value distribution applying to the smallest random time of failure out of many independent competing times. Consequently, one expects the Weibull will apply when failure occurs at a defect site within a material and there are many such sites competing with each other to be the first to produce a failure. Capacitor dielectric material is an example, and the Weibull distribution has proved very successful as a model for capacitor lifetime.

The best way to analyze Weibull data and estimate m and c is by the technique of maximum likelihood estimation (MLE). For censored and readout data, special computer programs are needed. A less optimal estimation procedure uses least squares and is equivalent to fitting a line to the data on Weibull graph paper using an objective regression procedure. Whatever method is used for the final parameter estimates, Weibull data plots, as shown in Chapter 6, should be part of the analysis.

5
The Normal and Lognormal Distributions

The lognormal distribution has become the most popular life distribution model for many high technology applications. In particular, it is very suitable for semiconductor degradation failure mechanisms. It has also been used successfully for modeling material fatigue failures and failures due to crack propagation. Some of its success comes from its theoretical properties; in other cases it "works" because it is flexible and fairly easy to use.

The primary purpose of this chapter is to discuss the properties and areas of application of the lognormal distribution. Many of these properties come directly from the properties of the normal distribution, because a simple logarithmic transformation changes lognormal data into normal data. Anything we know how to do with the normal distribution and normal data, we can therefore do for lognormal distribution and lognormal data. For that reason, we will start this chapter with a review of the normal distribution.

NORMAL DISTRIBUTION BASICS

Figure 5.1 shows the familiar bell-shaped curve that is the normal PDF. This curve is defined over the entire x axis, from $-\infty$ to $+\infty$. This domain of definition differs from the life distributions in the previous chapters, which were defined only on the positive x axis. The equation for the normal PDF is

$$f(x) = \frac{1}{\sigma\sqrt{2\pi}} e^{-(x-\mu)^2/2\sigma^2}$$

with the two parameters designated by μ (the mean), and σ (the standard deviation). As Figure 5.1 indicates, the distribution is symmetrical about

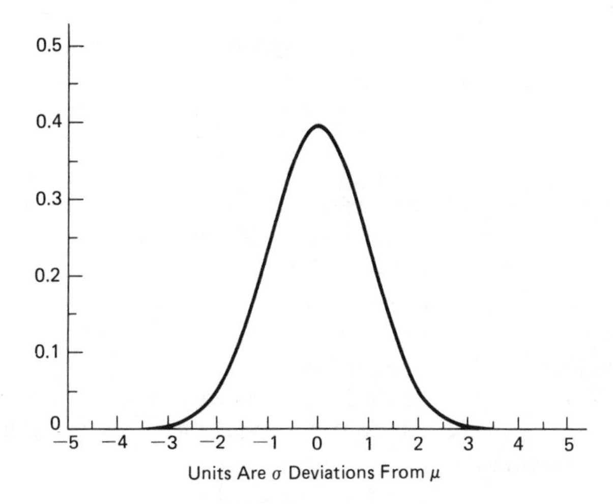

Figure 5.1. The Normal Distribution PDF.

its center μ and σ is a scale parameter which tells how close to the center the area under the curve is packed. A range of plus or minus one sigma from the center contains about 68% of the area or population; plus or minus two sigma contains about 95% and plus or minus three sigma covers 99.7% of any normal population.

The parameters μ and σ are natural in the sense that $\mu = E(X)$ or the mean, and $\sigma^2 = E[(X - \mu)^2]$ or the variance for a normal random variable X. The standard deviation, σ, is the square root of the variance.

The CDF for the normal distribution is obtained by integrating the PDF and has an S shape as shown in Figure 5.2. The hazard function is defined for all real x, but has limited value in a failure rate context, since the normal distribution is seldom used as a life distribution model.

If we start with a normal random variable X with parameters μ and σ, a simple transformation yields

$$Z = \frac{X - \mu}{\sigma}$$

This transformation is known as normalization, and Z has the "standard" normal distribution. This is a normal distribution with mean 0 and standard deviation 1. The symbols $\phi(z)$ and $\Phi(z)$ are commonly used to represent the standard normal PDF and CDF, respectively. The equations are

$$\phi(z) = \frac{1}{\sqrt{2\pi}} e^{-z^2/2}$$

$$\Phi(z) = \int_{-\infty}^{z} \phi(z)\,dz.$$

Since any normal variable can easily be transformed into a standard normal by subtracting its mean and dividing by its standard deviation, tables of $\Phi(z)$ have wide applicability. By symmetry, since $\Phi(z) = 1 - \Phi(-z)$, and $\Phi(0) = .5$, it is only necessary to table values of the CDF for positive z. Table 5.1 gives values of $\Phi(z)$ from $z = .01$ to $z = 3.50$, in steps of .01. This table will handle most applications.

Any area under the normal PDF curve, or any probability or population proportion, can be calculated from Table 5.1. An example will illustrate the techniques involved.

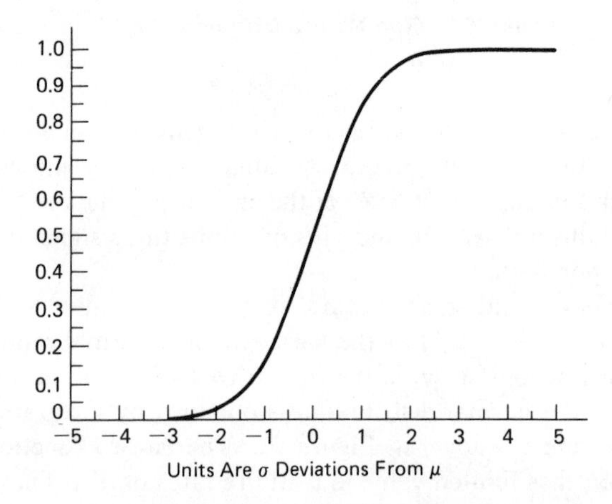

Figure 5.2. The Normal Distribution CDF.

EXAMPLE 5.1 NORMAL DISTRIBUTION CALCULATIONS

An electronics manufacturer uses interconnection wire which has a nominal strength of 11 g (i.e., it takes an average pull force of 11 g to break the wire). If the population distribution about this average value is normal with a standard deviation of 1.2 g, find the following:

a. The proportion of wires which will survive a pull of 13 g.
b. The probability a wire breaks under a load of 8 g.
c. The proportion of wires which will survive a pull of 8.5 g but not a pull of 13.2 g.

Table 5.1. Standard Normal CDF.

z	$\Phi(z)$	z	$\Phi(z)$	z	$\Phi(z)$	z	$\Phi(z)$	z	$\Phi(z)$
.00	.50000	.75	.77337	1.50	.93319	2.25	.98778	3.00	.99865
.01	.50399	.76	.77637	1.51	.93448	2.26	.98809	3.01	.99869
.02	.50798	.77	.77935	1.52	.93574	2.27	.98840	3.02	.99874
.03	.51197	.78	.78230	1.53	.93699	2.28	.98870	3.03	.99878
.04	.51595	.79	.78524	1.54	.93822	2.29	.98899	3.04	.99882
.05	.51994	.80	.78814	1.55	.93943	2.30	.98928	3.05	.99886
.06	.52392	.81	.79103	1.56	.94062	2.31	.98956	3.06	.99889
.07	.52790	.82	.79389	1.57	.94179	2.32	.98983	3.07	.99893
.08	.53188	.83	.79673	1.58	.94295	2.33	.99010	3.08	.99896
.09	.53586	.84	.79955	1.59	.94408	2.34	.99036	3.09	.99900
.10	.53983	.85	.80234	1.60	.94520	2.35	.99061	3.10	.99903
.11	.54380	.86	.80511	1.61	.94630	2.36	.99086	3.11	.99906
.12	.54776	.87	.80785	1.62	.94738	2.37	.99111	3.12	.99910
.13	.55172	.88	.81057	1.63	.94845	2.38	.99134	3.13	.99913
.14	.55567	.89	.81327	1.64	.94950	2.39	.99158	3.14	.99916
.15	.55962	.90	.81594	1.65	.95053	2.40	.99180	3.15	.99918
.16	.56356	.91	.81859	1.66	.95154	2.41	.99202	3.16	.99921
.17	.56750	.92	.82121	1.67	.95254	2.42	.99224	3.17	.99924
.18	.57142	.93	.82381	1.68	.95352	2.43	.99245	3.18	.99926
.19	.57535	.94	.82639	1.69	.95449	2.44	.99266	3.19	.99929
.20	.57926	.95	.82894	1.70	.95543	2.45	.99286	3.20	.99931
.21	.58317	.96	.83147	1.71	.95637	2.46	.99305	3.21	.99934
.22	.58706	.97	.83398	1.72	.95728	2.47	.99324	3.22	.99936
.23	.59095	.98	.83646	1.73	.95818	2.48	.99343	3.23	.99938
.24	.59483	.99	.63891	1.74	.95907	2.49	.99361	3.24	.99940
.25	.59871	1.00	.84134	1.75	.95994	2.50	.99379	3.25	.99942
.26	.60257	1.01	.84375	1.76	.96080	2.51	.99396	3.26	.99944
.27	.60642	1.02	.84614	1.77	.96164	2.52	.99413	3.27	.99946
.28	.61026	1.03	.84850	1.78	.96246	2.53	.99430	3.28	.99948
.29	.61409	1.04	.85083	1.79	.96327	2.54	.99446	3.29	.99950
.30	.61791	1.05	.85314	1.80	.96047	2.55	.99461	3.30	.99952
.31	.62172	1.06	.85543	1.81	.96485	2.56	.99477	3.31	.99953
.32	.62552	1.07	.85769	1.82	.96562	2.57	.99492	3.32	.99955
.33	.62930	1.08	.85993	1.83	.96638	2.58	.99506	3.33	.99957
.34	.63307	1.09	.86214	1.84	.96712	2.59	.99520	3.34	.99958
.35	.63683	1.10	.86433	1.85	.96784	2.60	.99534	3.35	.99960
.36	.64058	1.11	.86650	1.86	.96856	2.61	.99547	3.36	.99961
.37	.64431	1.12	.86864	1.87	.96926	2.62	.99560	3.37	.99962
.38	.64803	1.13	.87076	1.88	.96995	2.63	.99573	3.38	.99964
.39	.65173	1.14	.87286	1.89	.97062	2.64	.99585	3.39	.99965
.40	.65542	1.15	.87493	1.90	.97128	2.65	.99598	3.40	.99966
.41	.65910	1.16	.87698	1.91	.97193	2.66	.99609	3.41	.99968
.42	.66276	1.17	.87900	1.92	.97257	2.67	.99621	3.42	.99969
.43	.66640	1.18	.88100	1.93	.97320	2.68	.99632	3.43	.99970

Table 5.1.1 (*continued*)

z	$\Phi(z)$	z	$\Phi(z)$	z	$\Phi(z)$	z	$\Phi(z)$	z	$\Phi(z)$
.44	.67003	1.19	.88298	1.94	.97381	2.69	.99643	3.44	.99971
.45	.67364	1.20	.88493	1.95	.97441	2.70	.99653	3.45	.99972
.46	.67724	1.21	.88686	1.96	.97500	2.71	.99664	3.46	.99973
.47	.68082	1.22	.88877	1.97	.97558	2.72	.99674	3.47	.99974
.48	.68439	1.23	.89065	1.98	.97615	2.73	.99683	3.48	.99975
.49	.68793	1.24	.89251	1.99	.97670	2.74	.99693	3.49	.99976
.50	.69146	1.25	.89435	2.00	.97725	2.75	.99702	3.50	.99977
.51	.69497	1.26	.89617	2.01	.97778	2.76	.99711		
.52	.69847	1.27	.89796	2.02	.97831	2.77	.99720		
.53	.70194	1.28	.89973	2.03	.97882	2.78	.99728		
.54	.70540	1.29	.90147	2.04	.97932	2.79	.99736		
.55	.70884	1.30	.90320	2.05	.97982	2.80	.99744		
.56	.71226	1.31	.90490	2.06	.98030	2.81	.99752		
.57	.71566	1.32	.90658	2.07	.98077	2.82	.99760		
.58	.71904	1.33	.90824	2.08	.98124	2.83	.99767		
.59	.72240	1.34	.90988	2.09	.98169	2.84	.99774		
.60	.72575	1.35	.91149	2.10	.98214	2.85	.99781		
.61	.72907	1.36	.91308	2.11	.98257	2.86	.99788		
.62	.73237	1.37	.91466	2.12	.98300	2.87	.99795		
.63	.73565	1.38	.91621	2.13	.98341	2.88	.99801		
.64	.73891	1.39	.91774	2.14	.98382	2.89	.99807		
.65	.74215	1.40	.91924	2.15	98422	2.90	.99813		
.66	.74537	1.41	.92073	2.16	.98461	2.91	.99819		
.67	.74857	1.42	.92220	2.17	.98500	2.92	.99825		
.68	.75175	1.43	.92364	2.18	.98537	2.93	.99831		
.69	.75490	1.44	.92507	2.19	.98574	2.94	.99836		
.70	.75804	1.45	.92647	2.20	.98610	2.95	.99841		
.71	.76115	1.46	.92785	2.21	.98645	2.96	.99846		
.72	.76424	1.47	.92922	2.22	.98679	2.97	.99851		
.73	.76731	1.48	.93056	2.23	.98713	2.98	.99856		
.74	.77035	1.49	.93189	2.24	.98745	2.99	.99861		

$$\Phi(z) = \int_{-\infty}^{z} \phi(z)\,dz$$

SOLUTION

Before we can use tables based on the standard normal distribution, we have to normalize the numbers given in the problem. The proportion greater than 13 for a normal with $\mu = 11$ and $\sigma = 1.2$ is the same as the area to the right of $(13 - 11)/1.2 = 1.67$ for a standard normal distribution. This is $1 - \Phi\ (1.67)$. Table 5.1 gives $\Phi\ (1.67) = .95254$, so the answer is $1 - .95254$ or .04726.

The probability a wire breaks under an 8-g load is just $\Phi[(8 - 11)/1.2] = \Phi(-2.5)$. Since Table 5.1 has no negative z values, we use the symmetry relationship which says that $\Phi(-2.5) = 1 - \Phi(2.5)$. We look up $\Phi(2.5)$ in Table 5.1 and obtain $\Phi(-2.5) = 1 - .99379 = .00621$.

The final question asks for the area between the z values of $(8.5 - 11)/1.2$ and $(13.2 - 11)/1.2$. This region is the area to the left of 1.83, but to the right of −2.08. That area is given by $\Phi(1.83) - \Phi(-2.08)$. From Table 5.1, this proportion is $.96638 - (1 - .98124) = .94762$.

Some additional properties of normal random variables extend the usefulness of the standard normal CDF values given in Table 5.1 even further. These properties are:

1. If X and Y are independent normal random variables with X having mean μ_x and standard deviation σ_x, and Y having mean μ_y and standard deviation σ_y, then the sum $W = X + Y$ is also distributed normally with mean $\mu_w = \mu_x + \mu_y$ and standard deviation $\sigma_w = \sqrt{\sigma_x^2 + \sigma_y^2}$. This property can be extended to the sum of any number of independent normal variables by adding more mean terms to get the mean of the sum, and adding more variance terms under the square root sign to get the standard deviation of the sum.
2. Linear combinations of independent normal random variables are also normal. For X and Y as in property 1, $Z = aX + bY$ is normal with mean $(a\mu_x + b\mu_y)$ and standard deviation $\sqrt{a^2\sigma_x^2 + b^2\sigma_y^2}$.
3. If $x_1, x_2, \ldots, x_n$ are a sample of observations from a normal population with mean μ and standard deviation σ, then the sample mean $\bar{x} = \sum_{i=1}^{n} x_i/n$ also has a normal distribution with mean $\mu_{\bar{x}} = \mu$ and standard deviation $\sigma_{\bar{x}} = \sigma/\sqrt{n}$.

EXAMPLE 5.2 ROOT MEAN SQUARE EXAMPLE

A computer designer must carefully control the delay time it takes signals to travel from one location within a system to another, so that they arrive at their destination on time according to the system "clock." Each component within a given path has a nominal delay specified, as well as a three sigma upper and lower bound to use in testing the safety of the design. Assume a given path has four stages, two of which have nominal delays of 10 ns, with a three sigma upper bound of 13 ns, and the other two have nominal delays of 8 ns, with a three sigma upper bound of 11 ns. Characterize the nominal and three sigma upper bound of the total path and estimate the probability a delay is longer than 42.5 ns.

SOLUTION

The rules given for calculating the mean and standard deviation of a sum of independent normal random variables are valid even if the assumption of normality is dropped. If the delays at each stage are independent, the average or nominal total path delay is the sum of all the nominal delays (independence is not even required for this property to hold). Thus the nominal delay is $10 + 10 + 8 + 8 = 36$. The value of sigma for the total delay (here we do require independence) is the square root of the sum of all the individual variances. Three sigma for the first two stages is $13 - 10 = 3$, so sigma for each of these is 1. Three sigma for the next two stages is $11 - 8 = 3$, and again sigma is 1. The total path sigma is the square root of $1^2 + 1^2 + 1^2 + 1^2$, or $\sqrt{4} = 2$.

Since normality is usually a reasonable assumption for delay lengths, the probability of any delay exceeding 42.5 ns can be calculated by finding $1 - \Phi[(42.5 - 36)/2]$. Table 5.1 gives .99942 for $\Phi(3.25)$, so the probability of a delay this large is .00058.

Note that adding up three sigma limits for each stage (i.e., adding $13 + 13 + 11 + 11 = 48$) would give a six sigma upper limit on the path delay ($36 + 6 \times 2 = 48$). This would be an unrealistic limit.

APPLICATIONS OF THE NORMAL DISTRIBUTION

As the last example showed, the normal distribution is frequently used to model measurement errors of almost any kind. In addition, many of the populations frequently encountered in industrial (or other) applications have bell-shaped symmetrical histograms that can be fit adequately by the normal PDF.

Even if a population has a skewed histogram that cannot be modeled by a normal distribution, the mean of a sample from this population (or $\bar{x}$), considered as a random variable in its own right, will tend to have a more nearly normal shape. (This property is a consequence of the central limit theorem, described in the next section.) Therefore, we can use the normal distribution to model populations of sample means, with nearly universal success. This feature may not seem very useful, but in fact, it is the basis for the wide applicability of control chart techniques for monitoring industrial processes.

For any process we want to maintain at its best operating level, we can take regular samples of a key process parameter and treat the sample averages as if they were normally distributed. A control chart has upper and lower lines located three sigma units (sigma in terms of the sample averages, or $\sigma/\sqrt{n}$, if σ is the population standard deviation and n is the sample size) above and below the overall process parameter average. Based on properties of the normal distribution, only about 3 sample averages in 1000 should fall outside of the region bounded by the three sigma lines. In other words, plus or minus three sigma limits from the center of the population contain 99.7% of the population values, as can be calculated from Table 5.1 by looking at the area to the left of 3 and to the right of −3. So any point that does fall outside is highly unusual and unlikely, unless something abnormal happened to change the process. Control charts are discussed in more detail in Chapter 9.

If we investigate the process every time an "out of control point" outside of the control limits occurs, we will be chasing false alarms only an average of 3 times in 1000, or once every six and a half years if we take samples every week. This rare false alarm is a small price to pay for an alert system that tells us some probably unusual variability has been introduced into our process.

THE CENTRAL LIMIT THEOREM

The central limit theorem, or CLT, which ensures the approximate normality of $\bar{x}$, has many versions. Most statements involve intricate mathematical conditions (see Gnedenko et al., 1969 page 33 for example). For our purposes, a heuristic verbal statement will suffice.

Under fairly general conditions, the CLT shows that the sum of a large number of random variables, each of which contributes only a small amount to the total, will have an approximately normal distribution. As the number of contributing factors increases, with the share of each growing smaller, the approximation becomes more exact.

The sample average, as the sample size n increases, is a perfect example of a situation satisfying the conditions of the CLT. Each term in the sum $\bar{x} = (x_1/n) + (x_2/n) + \cdots + (x_n/n)$ grows smaller as n increases, and the distribution of $\bar{x}$ approaches normality no matter what underlying distribution the x_i come from (as long as this distribution has a finite mean and variance). In terms of practical application, when n is as little as 4 or 5, $\bar{x}$ typically shows good normal characteristics.

Another excellent example is the variability introduced whenever anything is measured. Measurement error comes from a multiplicity of small random factors, combining to produce a deviation from the "true" value of the quantity under measurement. Repeated measurements produce a histogram with a normal shape, as one would expect from the CLT. In fact, the great nineteenth century mathematician Karl Gauss derived the normal distribution as the appropriate model for measurement errors, and it is often called the Gaussian error distribution.

An example similar to measurement error would be the variability when a mechanical operation is repeated over and over again. (This formed the basis of Example 4.3 of the preceding chapter, where we saw that normal errors in two coordinate directions led to a Rayleigh distribution for the radial error.) The rule of thumb of designing so that three sigma deviations from nominal are within specifications comes from assuming that manufacturing variation and repeatability variation leads to a normal shaped variability about the nominal value, and plus or minus three sigma limits will therefore capture 99.7% of the population.

The CLT justifies the empirical observation that the normal model works for many types of data—it makes sense to model many important random phenomena as the sum of a large number of small contributing factors. Since fitting a normal model to data means picking a $\hat{\mu}$ estimate of μ and a $\hat{\sigma}$ estimate of σ, methods of estimation will be looked at next.

NORMAL DISTRIBUTION PARAMETER ESTIMATION

The standard estimates derived from complete samples of observations from a normal distribution are well known and the best that can be used. For a sample of size n, they are

$$\hat{\mu} = \bar{x} = \sum_{i=1}^{n} \frac{x_i}{n}, \qquad \hat{\sigma} = s = \sqrt{\sum_{i=1}^{n} \frac{(x_i - \bar{x})^2}{n-1}}.$$

These are also the MLEs, except that the MLE of σ has a divisor of n, instead of $(n - 1)$, which is a negligible difference for large sample sizes.

If the data are censored or multicensored or grouped by readout intervals, the standard estimates no longer apply. Either graphical estimation (see Chapter 6) or computer programs using the maximum likelihood method can be used. There is also a least-squares procedure equivalent to fitting a straight line through the data points plotted on normal probability paper, which can be used to calculate μ and σ estimates.

The least-squares approach starts by setting the CDF estimate at any of the data points, or readout times, approximately equal to the standard normal CDF Φ evaluated at the appropriate normalized point. The inverse of Φ is then applied to derive an equation that is linear in the unknown parameters μ and σ. The equations are

$$\Phi\left(\frac{x-\mu}{\sigma}\right)=F(x)$$

$$\Phi^{-1}\Phi\left(\frac{x-\mu}{\sigma}\right)=\Phi^{-1}F(x)$$

$$\frac{x-\mu}{\sigma}=\Phi^{-1}F(x)$$

$$x=\mu+\sigma\Phi^{-1}F(x).$$

The last equation is the key one, since it links an observable variable x to a quantity $\Phi^{-1}F(x)$ that can be calculated from the CDF estimate corresponding to x. Indeed, this linear equation is the basis of probability plotting, discussed in Chapter 6.

For r exact observations censored out of a sample of n, first order the points so that x_i is the ith smallest (x_1 is the smallest), and then estimate the CDF $F(x_i)$ by $(i - .3)/(n - .4)$. If the data come in grouped intervals with endpoints $x_1, x_2, \ldots, x_k$, then estimate $F(x_i)$ by the number of observations less than or equal to x_i divided by n. (The rationale for these CDF estimates is given in Chapter 6.)

The quantity $\Phi^{-1}(F)$ can be obtained from Table 5.1. For example, if $F = .62$, look up this value on the probability scale and find that the z value that has $\Phi(z) = .62$ is approximately .31. Therefore, $\Phi^{-1}(.62) = .31$. Since Table 5.1 only has probability values $\geq .5$, we have to use the symmetry relationship $\Phi^{-1}(F) = -\Phi^{-1}(1 - F)$ for any F that is less than .5. If we want $\Phi^{-1}(.33)$, we look in Table 5.1 for a probability of $1 - .33 = .67$, and obtain $z = .44$. This shows $\Phi^{-1}(.33) = -.44$.

EXAMPLE 5.3 CENSORED NORMAL DATA

The electronics manufacturer in Example 5.1 is considering buying interconnection wires from a new vendor. A reliability engineer is asked to estimate the pull strength distribution from a sample of 500 wires. His test apparatus increases the number of grams of pull in steps of 1, from 1 g to 13. His procedure is to test each wire and record at what step it breaks, or whether it went all the way up to a pull of 13 g without breaking. He observes 1 failure at 9 g, 10 at 10 g, 79 at 11 g, 164 at 12 g, and 170 at 13 g. There were 13 wires that survived the test. Calculate least-squares estimates of μ and σ.

SOLUTION

The data are grouped by readout intervals so the proper CDF estimate to use at a pull strength step is the total number of failures up to that point

Table 5.2. Example 5.3 Worksheet.

READOUT POINT	CDF ESTIMATE F	$\Phi^{-1}(F)$
9	1/500 = .002	−2.88
10	11/500 = .022	−2.01
11	90/500 = .180	−0.92
12	254/500 = .508	.02
13	424/500 = .848	1.03

divided by the sample size of 500. Table 5.2 shows a worksheet for setting up the inputs to a least-squares routine. The inverse CDF column was calculated using Table 5.1.

Inputting the first column into a regression program as the "Y" or dependent variable, and inputting the third column as the "X" or independent variable yields an intercept estimate of $\hat{\mu} = 11.97$, and a slope estimate of $\hat{\sigma} = 1.01$. (A maximum likelihood method program for readout data gives exactly the same estimates.)

This survey of some of the properties of the normal distribution has been included primarily for what will be needed when we discuss the lognormal life distribution. Major topics, such as confidence intervals for parameter estimates and tests of hypotheses concerning normal populations, have not even been mentioned. The reader interested in reviewing these areas of basic statistics should consult a text such as Dixon and Massey (1969) or Ostle and Mensing (1975).

THE LOGNORMAL LIFE DISTRIBUTION

The simple relationship between a normal random variable X, and its derived lognormal random variable t_f is the following: if X has mean μ and standard deviation σ, then

$$t_f = e^X$$

has a lognormal distribution with parameters $T_{50} = e^{\mu}$ and σ. Alternatively, if we start with a population of random failure times t_f modeled by a lognormal distribution with median parameter T_{50} and shape parameter σ, then the population of logarithmic failure times $X = \ln t_f$ is normal with mean $\mu = \ln T_{50}$ and standard deviation σ.

The logarithm of a lognormal is normal. In a mathematical sense, we never have to deal with the lognormal as a separate distribution; we can take logarithms (natural) of all the data points and analyze the transformed data as we would analyze normal data. This procedure is the basis for almost all lognormal analysis routines. After completing a normal analysis on the logarithmic time scale, the results are displayed in terms of the lognormal distribution and real time.

The T_{50} parameter of the lognormal distribution is a "natural" parameter in the sense that it is the median time to fail of the population of lognormal lifetimes. The parameter σ, on the other hand, causes much confusion, as it is not a "natural" quantity describing the population of times to fail. σ *should be thought of only as a shape parameter for the lognormal distribution. It is not the standard deviation of the population of lifetimes.* σ is a standard deviation, in units of logarithmic time, for the normal distribution describing the population of logarithmic times to failure. In a sense, σ is a "borrowed" parameter, only used for the lognormal distribution because of mathematical convenience.

PROPERTIES OF THE LOGNORMAL DISTRIBUTION

The PDF for the lognormal distribution is given by

$$f(t) = \frac{1}{\sigma t \sqrt{2\pi}} e^{-(1/2\sigma^2)(\ln t - \ln T_{50})^2}.$$

The CDF $F(t)$ is the integral of the PDF from 0 to time t. It can also be expressed in terms of the standard normal CDF as

$$F(t) = \Phi\left\{\frac{\ln(t/T_{50})}{\sigma}\right\}.$$

This last equation allows us to use Table 5.1 to evaluate probabilities of failure and survival for the lognormal distribution.

EXAMPLE 5.4 LOGNORMAL PROBABILITIES

A population of components, when tested at high laboratory stresses, fail according to a lognormal distribution with $T_{50} = 5000$ and $\sigma = .7$. What percent of failures are expected on a 2000-hr test?

SOLUTION

Direct substitution gives

$$F(2000) = \Phi\left\{\frac{\ln(2000/5000)}{.7}\right\} = \Phi(-1.309) = 1 - \Phi(1.309).$$

Using Table 5.1, we get 1 − .905 = .095 or 9.5%.

The lognormal PDF has a wide variety of appearances, depending on the critical shape parameter σ. As shown in Figure 5.3, the variety of shapes lognormal data can have resembles the shapes taken on by the Weibull distribution (see Chapter 4, Figure 4.1). This flexibility makes the lognormal an empirically useful model for right skewed data. The similarity between Figure 5.3 and the Weibull PDF shapes shown in Figure 4.1 also suggests that both models will often fit the same set of experimental data equally well. One trick that occasionally helps choose whether a lognormal or a Weibull will work better for a given set of data is to look at a histogram of the logarithm of the data. If this is symmetrical and bell shaped, the lognormal will fit the original data well. If, on the other hand, the histogram now has a left skewed appearance, a Weibull fit to the original data might work better.

Some examples of lognormal CDF curves are plotted in Figure 5.4. These curves show a long right-hand tail (right skew) for large values of σ. For a small σ such as .2, the PDF and CDF have a normal shape.

The lognormal failure rate function $h(t)$ has to be calculated using the basic definition $h(t) = f(t)/[1 - F(t)]$.

The various shapes that can be taken by the lognormal failure rate are shown in Figure 5.5. These are also similar to the variety of shapes taken

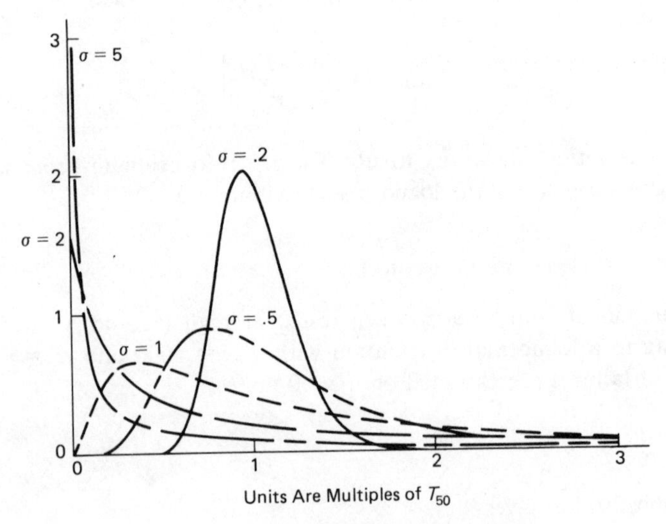

Figure 5.3. Lognormal Distribution PDF.

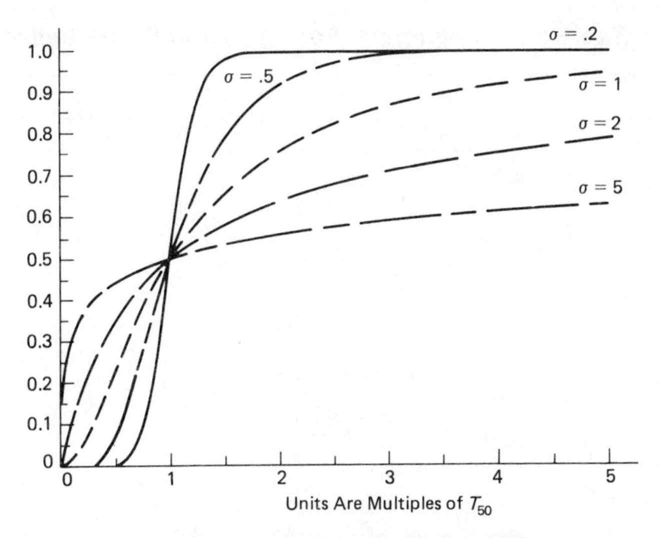

Figure 5.4. Lognormal Distribution CDF.

on by the Weibull failure rate (as shown in Figure 4.2). Large values of σ behave like small values of m for the Weibull; if σ is larger than 2, the failure rate has a decreasing shape (although it always starts at 0 and rises to a maximum before decreasing—for large σ the rise is so quick that, for practical purposes, the failure rate appears to decrease throughout life). Values of σ around 1 give failure rate functions that first rise quickly, then have a

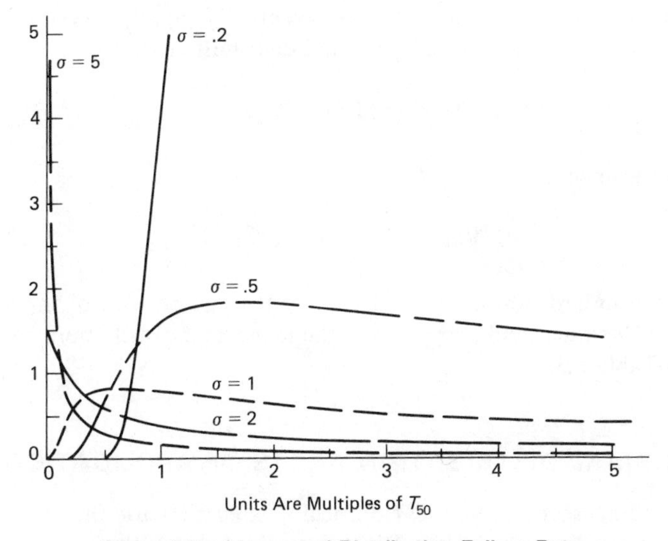

Figure 5.5. Lognormal Distribution Failure Rate.

Table 5.3. Lognormal Formulas and Properties.

NAME	FORMULA OR PROPERTY
PDF	$f(t) = \frac{1}{\sigma t \sqrt{2\pi}} e^{-(1/2\sigma^2)(\ln t - \ln T_{50})^2}$
CDF	$F(t) = \int_0^t f(t)\,dt$ $= \Phi \left\{ \frac{\ln t / T_{50}}{\sigma} \right\}$
Reliability	$R(t) = 1 - F(t)$
Failure rate	$h(t) = \frac{f(t)}{R(t)}$
T_{50}	Median lifetime or 50% failure point
σ or sigma	Shape parameter. Large σ (≥ 2) means high early failure rate decreasing with time. Low σ ($\leq .5$) means increasing (wearout) type failure rate and a PDF with a "normal" shape. For σ close to 1, the failure rate is fairly flat.
Relation to normal	If t_f is lognormal with parameters (T_{50}, σ), then $X = \ln t_f$ is normal with mean $\mu = \ln T_{50}$ and standard deviation σ.
Mean	$E(t) = T_{50}\, e^{\sigma^2/2}$
Variance	$\text{Var}(t) = T_{50}\, e^{\sigma^2} (e^{\sigma^2} - 1)$

fairly constant failure rate for a long period of time. Low values of σ correspond to wearout failure rates that are eventually rapidly increasing.

The mean, or MTTF, for a lognormal distribution is given by

$$E(t_f) = \text{MTTF} = T_{50}\, e^{\sigma^2/2}$$

and the variance is

$$\text{Var}\,(t_f) = T_{50}\, e^{\sigma^2}(e^{\sigma^2} - 1).$$

The true standard deviation is, of course, the square root of the variance.

The key formulas and properties of the lognormal distribution are summarized in Table 5.3.

LOGNORMAL DISTRIBUTION AREAS OF APPLICATION

The preceding section showed empirical justification for the suitability of the lognormal distribution as a life distribution model: the lognormal has a very flexible PDF and failure rate function, and its relationship to the normal distribution makes it convenient to work with.

Even though both the lognormal model and the Weibull model will fit most sets of life test data equally well, there is a major difference between them that can have critical importance. That difference comes about when we use the fitted model to extrapolate beyond the range of the sample data.

If we have 100 units on test, the smallest percentile we can actually observe is the 1% point. We need a parametric model to project to the earlier time when .001% of the population might fail, or to estimate a proportion of failures for a time much smaller than the time of the first observed test failure. If we use a lognormal model, the projection to smaller percentiles is usually optimistic, as compared to a projection based on a Weibull distribution fit. In other words, the lognormal will extrapolate lower AFRs at early times. Sometimes the difference can be several orders of magnitude.

In Chapter 7 we will see that the use failure rates we are concerned with often must be estimated from the very early percentiles of a laboratory test conducted at high stress. How, then, can we decide whether it is appropriate to use the more optimistic lognormal model or the Weibull model?

A reasonable answer to this is to look for a theoretical justification for one model or the other, based on the failure mechanism under investigation. In Chapter 4 we learned that the Weibull can be derived as the extreme value distribution that applies when many small defect sites compete with each other to be the one that causes the earliest time of failure. Similarly, there is a derivation that leads to the lognormal as a model for processes that degrade over time, eventually reaching a failure state.

The precise model that leads to a lognormal distribution is called a multiplicative (or proportional) growth model. At any instant of time, the process undergoes a random increase of degradation that is proportional to its present state. The multiplicative effect of all these random (and assumed independent) growths builds up to failure.

This model was used with great success by Kolmolgorov (1941) to describe the dimensions of particles (such as grains of sand by the ocean) constantly undergoing a pulverizing process. References describing the multiplicative model include Mann et al. (1974) and Gnedenko et al. (1969).

The derivation depends on viewing the process at many discrete points of time over the interval $(0, t)$. Let $x_1, x_2, x_3, \ldots, x_n$, be random variables measuring the state of degradation of the process as we move along in time. The multiplicative model says that

$$x_i = (1 + \delta_i)x_{i-1}$$

where δ_i is the small proportional growth that takes the process from x_{i-1} to x_i. Then it follows that

$$x_n = \left[\prod_{i=1}^{n} (1 + \delta_i) \right] x_0$$

$$\ln x_n = \sum_{i=1}^{n} \ln(1 + \delta_i) + \ln x_0$$

$$\ln x_n \approx \sum_{i=1}^{n} \delta_i + \ln x_0.$$

Now we invoke the central limit theorem on the sum of the small random quantities δ_i, and obtain that $\ln x_n$ has an approximate normal distribution. Therefore, x_n, and hence x_t, has a lognormal distribution. This argument shows that the probability the process has degraded to a failure state by time t is approximately given by the lognormal distribution.

This derivation gives us a theoretical reason to prefer a lognormal model when we can hypothesize a multiplicative degradation process is going on. Many semiconductor failure mechanisms could fit this model. Some examples are chemical reactions such as corrosion, or material movement because of diffusion or migration or even crack growth propagation. It makes more sense to use a lognormal model for these kinds of failures, than a Weibull (extreme value) model.

LOGNORMAL PARAMETER ESTIMATION

If we have a complete sample of exact times to failure, then the best way to estimate T_{50} and σ is to take natural logarithms of all times of failure and then calculate $\bar{x}$ and s for the sample of logarithmic data. The estimate of T_{50} is $e^{\bar{x}}$, and the estimate of σ is s. The formulas are

$$\bar{x} = \frac{\sum_{i=1}^{n} \ln t_i}{n}$$

$$s = \sqrt{\frac{\sum_{i=1}^{n} (\ln t_i - \bar{x})^2}{n - 1}}$$

$$\hat{T}_{50} = e^{\bar{x}}$$

$$\hat{\sigma} = s.$$

If the data are censored or grouped by readout intervals, these estimates cannot be calculated. Instead, a computer program that çalculates MLEs should be used (see the description of MLEs in Chapter 4). Commercial programs like Statpac (see Strauss, 1980) and Censor (see Meeker and Duke, (1981) are available.

Graphical methods can also be used to obtain (less precise) estimates. These procedures are described in Chapter 6. Here we will show a least-squares technique for putting a straight line through the data points one would plot on lognormal probability paper. This is an analytic method for estimating lognormal parameters that can be used for censored or readout data. The estimates do not have the large-sample optimality properties of MLEs, but when MLE computer programs are not available, it is a reasonable method to use.

The least-squares technique is similar to that described earlier for the normal distribution. By using the relationship between the lognormal and the normal, the linear equation involving the estimated sample CDF given in the section on normal parameter estimation becomes

$$\ln t_f = \ln T_{50} + \sigma \Phi^{-1} F(t).$$

Least-squares estimates of $\ln T_{50}$ and σ are obtained from any standard regression program by inputting the $\ln t_i$ values as the dependent or Y variable, and the CDF estimates as the independent or X variable (similar to Example 5.3). After the least-squares fit is completed, the slope estimate is $\hat{\sigma}$ and the antilogarithm of the intercept is $\hat{T}_{50}$.

EXAMPLE 5.5 LOGNORMAL PARAMETER ESTIMATION

A 4000-hr life test of 20 components yields 9 failures. Table 5.4 shows the exact times of failure data, along with the corresponding CDF estimates (the formula used for the CDF estimates when exact times of failure are recorded is explained in Chapter 6). Estimate T_{50} and σ for this data.

SOLUTION

The natural logarithm of the time of failure column is the dependent variable, and the standard normal inverse of the CDF column is the independent variable. The dependent variable values are: 7.183, 7.716, 7.718, 7.733, 7.767, 7.928, 7.936, 8.04, 8.188. The independent variable values are: −1.825, −1.385, −1.117, −.911, −.739, −.585, −.445, −.313, −.186. Least-squares estimates of the intercept and slope are 8.219 and .5, respectively. The T_{50} estimate is $e^{\text{intercept}}$ or 3710. The sigma estimate is the slope or $\hat{\sigma} = .5$. MLEs for the same data are $\hat{T}_{50} = 4164$ and $\hat{\sigma} = .58$. (This example used simulated data from a population with true $T_{50} = 5000$ and true $\sigma = .7$, so in this example, the MLEs were more accurate.)

Table 5.4. Life Test Failure Data (20 Units on Test).

ORDER OF FAILURE	TIME OF FAILURE	CDF ESTIMATE $(i - .3)/20.4$
1	1317	.034
2	2243	.083
3	2248	.132
4	2282	.181
5	2362	.230
6	2773	.279
7	2797	.328
8	3104	.377
9	3600	.426

The equation used for least-squares parameter estimates can also be solved for T_{50}, assuming σ and $F(t)$ are known, or solved for σ, assuming T_{50} and $F(t)$ are known, or even solved for t, assuming the other quantities are known. These alternatives are often useful, and the equations are

$$T_{50} = te^{-\sigma\Phi^{-1}F(t)}$$

$$\sigma = \frac{\ln(t/T_{50})}{\Phi^{-1}F(t)}$$

$$t = T_{50}\, e^{\sigma\Phi^{-1}F(t)}.$$

EXAMPLE 5.6

What must the T_{50} of a lognormal distribution be in order to have .1% cumulative failures by 60,000 hr, assuming σ is known to be .8? What σ is needed in order to have an AFR over the first 40,000 hr of life of .05%/K, given the T_{50} is 300,000? For this σ, how many hours does it take to reach 1% cumulative failures?

SOLUTION

We have

$$T_{50} = 60{,}000e^{-.8\Phi^{-1}(.001)}$$

and we use Table 5.1 to find $\Phi^{-1}(.001)$ by finding $-\Phi^{-1}(.999)$. We get -3.09 and $T_{50} = 60{,}000e^{2.472} = 710{,}767$.

To answer the second question, we first have to convert the AFR given to a CDF value. But $\text{AFR}(40{,}000) = 10^5 \times -\ln R(40{,}000)/40{,}000$, where the factor of 10^5 is needed to convert the units to %/K. Setting this AFR equal to .05, we obtain $R(40{,}000) = .9802$ and $F(40{,}000) = .0198$. Using the formula $\sigma = \ln(40{,}000/300{,}000)/\Phi^{-1}(.0198)$, and the Table 5.1 value for $\Phi^{-1}(.0198) = -\Phi^{-1}(.9802) = -2.06$, we obtain $\sigma = .98$.

The time point when the CDF is 1% is given by

$$t = 300{,}000e^{.98\Phi^{-1}(.01)} = 300{,}000e^{-.98 \times 2.33} = 30{,}581.$$

SUMMARY

The normal distribution, while not a suitable model for population lifetimes, finds many applications in reliability analysis because of its relationship to the lognormal distribution and its use as an error model and a model for control chart applications (see Chapter 9). Since any normal variable can be transformed into a standard normal variable by subtracting its mean and dividing by its standard deviation, the standard normal table of CDF values given in Table 5.1 are frequently used.

Estimates of the normal distribution parameters μ and σ are $\bar{x}$ and s for complete samples. For censored and grouped data, graphical methods (Chapter 6), or the method of maximum likelihood, can be used. A simple regression procedure that objectively finds the best line to fit the data points on normal probability paper, can also be used to obtain estimates.

The lognormal distribution is a very useful and flexible model for reliability data. It is closely related to the normal distribution since the logarithm of a lognormal random variable has a normal distribution. The parameters of the lognormal distribution are its median, T_{50}, and a shape parameter σ. If we work in a logarithmic time scale, by taking natural logarithms of all the data points, the resulting normal population has mean parameter $\mu = e^{T_{50}}$ and standard deviation σ. This feature allows us to use Table 5.1 to calculate probabilities associated with a lognormal distribution.

The lognormal distribution can be derived from a model for degradation processes. The main requirement is that the change in the degradation process at any time be a small random proportion of the accumulated degradation up to that time. This derivation may explain why the lognormal has been so successful modeling failures due to chemical reactions or molecular diffusion or migration. Some types of crack growth might also be expected to have a lognormal distribution.

Formulas and properties associated with the lognormal distribution are shown in Table 5.3. Estimation methods are identical to those mentioned for the normal distribution, because of the close relationship between the lognormal and the normal.

6
Reliability Data Plotting

We have discussed in previous chapters analytical techniques for various distributions. However, it is useful to have simple graphical procedures that allow checking the applicability or quickly estimating certain parameters of the assumed distribution model. The methods we show in this chapter will permit rapid analysis of both exact time of failure data or interval, that is, readout type data. Complete, single censored, or multicensored data will be treated. We will illustrate the procedures for the exponential, Weibull, normal, and lognormal distributions. We begin this chapter by first reviewing the properties of straight lines. Then the concepts of least-squares regression analysis, and of linear rectification are described. This introductory material will prepare us for the graphical examples that follow.

PROPERTIES OF STRAIGHT LINES

Consider the straight line drawn through two points P_1 at (x_1,y_1) and P_2 at (x_2,y_2) in Figure 6.1. The slope of the line is defined as the ratio of the change in the y coordinates to the change in the x coordinates, for points on the line. Designating the slope by the letter m, we have

$$m = \frac{\text{rise}}{\text{run}} = \frac{y_2 - y_1}{x_2 - x_1} = \frac{\Delta y}{\Delta x}$$

Note the calculation of the slope is independent of the location of the points on the line, but the best precision will be obtained by selecting the two points as far apart as possible. We can write the above equation in the form

$$\Delta y = m\,\Delta x,$$

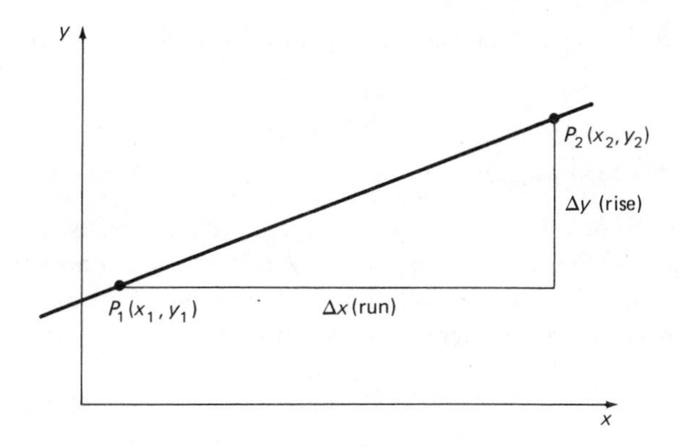

Figure 6.1. Straight line plot.

which says that the change in y is proportional to the change in x, and the slope m is the proportionality factor. This property is called a linear relation.

Parallel lines have equal slopes. It is easy to show (Thomas, 1960) that perpendicular lines have the product of the slopes equal to -1.

Consider the expression

$$y_2 - y_1 = m(x_2 - x_1).$$

Fix the point (x_1, y_1) on the line. For any (x, y) on the line,

$$y - y_1 = m(x - x_1),$$

or

$$y = mx + (y_1 - mx_1).$$

Since $y_1 - mx_1$ is fixed, we can set this quantity equal to a constant b and write

$$y = mx + b.$$

b is called the intercept since at $x = 0$, $y = b$, indicating that b is the y coordinate at the point where the straight line crosses the y axis.

An alternative expression for a straight line is

$$Ax + By + C = 0,$$

where A, B, C are constants. Any equation in this form is an equation for a straight line.

EXAMPLE 6.1 LINEAR EQUATIONS

Let C and F denote, respectively, corresponding Celsius and Fahrenheit temperature readings. Given that there is a linear relation between the two readings, F and C, find the equation from the data $F = 32$ at $C = 0$, and $F = 212$ at $C = 100$. Is there a temperature at which $F = C$?

SOLUTION

Use

$$F_2 - F_1 = m(C_2 - C_1).$$

Then,

$$212 - 32 = m(100 - 0),$$

or

$$m = \frac{180}{100} = \frac{9}{5}.$$

At $C = 0$, $F = b = 32$. So the equation is $F = (9/5)C + 32$. If we want the temperature at which $F = C$, we solve $F = (9/5)F + 32$, or $F = -40$.

LEAST-SQUARES FIT (REGRESSION ANALYSIS)

Because of variation that will occur whenever measurements are made in experimentation, even if a linear relationship is expected between two variables, there will be a scatter of data points around the expected line.

Is there a "best fitting" line for a given set of data points? What do we mean by best fit? Intuitively, we would like to be able to accurately predict the y value from a given x value. We assume y is the dependent variable and x is the independent variable. Then, one definition of best fit would be to choose a line such that the sum of the squares of the deviations of the predicted y values from the observed y values is a minimum. In statistics such a line is called a regression line (see Figure 6.2).

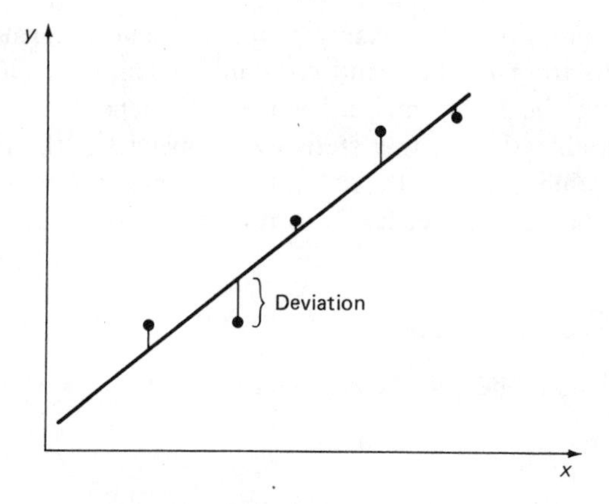

Figure 6.2. Regression Line Example.

It can be shown (Mendenhall and Scheaffer, 1973) that the equation of the best fitting line is

$$\hat{y} = mx + (\bar{y} - m\bar{x}),$$

where

$$\bar{x} = \text{mean of } x = \frac{\sum_{i=1}^{n} x_i}{n}; \qquad \bar{y} = \text{mean of } y = \frac{\sum_{i=1}^{n} y_i}{n}.$$

and the slope of the regression line is

$$m = \frac{n \sum_{i=1}^{n} x_i y_i - \left(\sum_{i=1}^{n} x_i\right)\left(\sum_{i=1}^{n} y_i\right)}{n \sum_{i=1}^{n} x_i^2 - \left(\sum_{i=1}^{n} x_i\right)^2}.$$

The intercept is $b = \bar{y} - m\bar{x}$. Note when $x = \bar{x}$, we have $y = \bar{y}$, that is, the line goes through the point $(\bar{x},\bar{y})$.

While this formula appears a bit tedious to evaluate, it is expected that most individuals working in the field of reliability will have access to statistical packages that will perform the calculations. Even most inexpensive hand calculators are designed to do least-squares regression calculations.

If certain assumptions (for example, independence, normal distribution of the deviations from the line, and constant variance for all x values) are made concerning the linear model relating the dependent variable to the independent variable, then further statistical properties of the slope, intercept, and predicted values can be derived. The reader is referred to a text like that by Mendenhall and Scheaffer for a full discussion of this topic.

EXAMPLE 6.2 REGRESSION LINE

Find the regression line for the x,y pairs: (1,3.1), (2,4.0), (3,5.1), (4,5.7), and (5,7.1).

SOLUTION

Set up simple table.

	x	y	x^2	xy
	1	3.1	1	3.1
	2	4.0	4	8.0
	3	5.1	9	15.3
	4	5.7	16	22.8
	5	7.1	25	35.5
Sums	15	25.0	55	84.7

$$\bar{x} = 3 \qquad \bar{y} = 5 \qquad \left(\sum_{i=1}^{n} x_i\right)^2 = 225$$

$$m = \frac{5 \times 84.7 - 15(25.0)}{5 \times 55 - 225} = 0.97 \quad \text{and} \quad b = \bar{y} - m\bar{x} = 2.09.$$

Thus, $\hat{y} = 0.97x + 2.09$.

RECTIFICATION

The techniques of linear regression can be used to estimate the constants and parameters of many equations, either directly or via a procedure called linear rectification. The simplest example is the problem of determining a resistance value from current and voltage pairs (I, V). By Ohm's law, $V =$

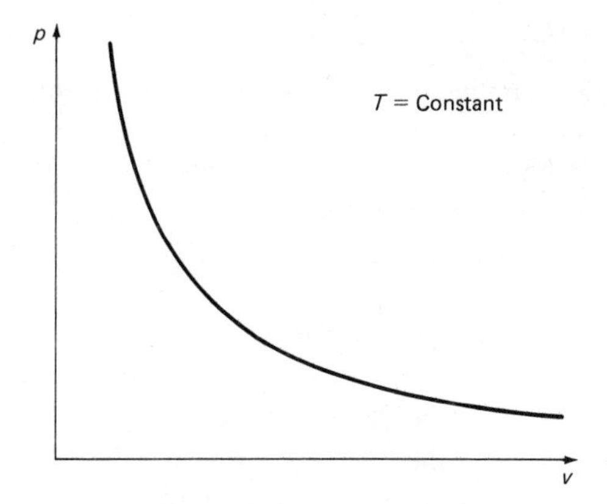

Figure 6.3. Gas Law Plot.

IR, where I is the current, V is the voltage, and R is the resistance. This equation is already in the form of $y = mx + b$, with $y = V$, $x = I$, $m = R$, and $b = 0$. So a plot of V versus I should approximate a straight line, going through the origin (0,0), with slope R.

Consider the gas law given by $pv = RT$, where p is the pressure, v is the volume, T is the temperature, and R is the gas constant. Suppose we wish to determine R from the experiment in which the volume is varied at a constant temperature (adiabatically) and the pressure is measured. A plot of p versus v would appear as Figure 6.3. However, if p is plotted versus the reciprocal of v, that is, $y = p$, $x = 1/v$, then we should obtain a straight line with slope RT. Since T is known and fixed, R is simply determined. See Figure 6.4.

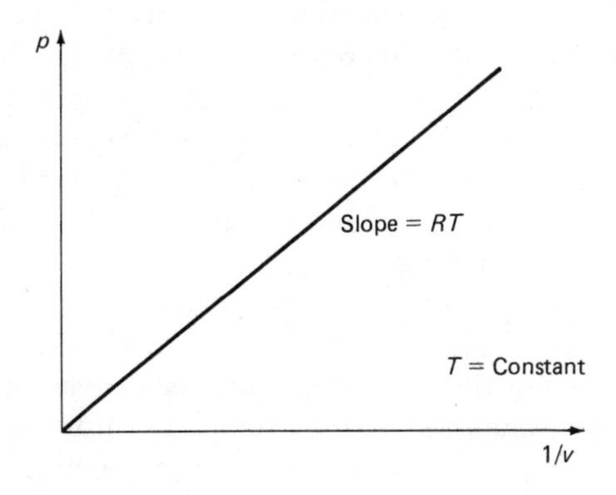

Figure 6.4 Gas Law Plot Using Rectification.

EXAMPLE 6.3 LINEAR RECTIFICATION

A model states the variable Q is proportional to some power of the variable s, that is, $Q = \alpha s^k$, where k is unknown and α is the unknown constant of proportionality. How do we determine k and α?

SOLUTION

Taking logs of both sides of the equation, we get

$$\ln Q = \ln \alpha + \ln s^k$$
$$\ln Q = \ln \alpha + k \ln s.$$

If we now plot Q versus s on log–log paper (or plot $\ln Q$ versus $\ln s$ on linear–linear paper), we should get a straight line with slope k and intercept α (or $\ln \alpha$ on linear–linear paper). Note on log–log paper, the intercept is read at $s = 1$, since $\ln 1 = 0$.

The concept of linear rectification (or doing what is necessary to convert an equation into a linear form) is very important because it may allow us to determine easily estimates of the parameters of a model from experiments. This utility shall become very evident in Chapter 7 when acceleration models are introduced. Now, however, we shall see the application of rectification to probability plotting.

PROBABILITY PLOTTING FOR THE EXPONENTIAL DISTRIBUTION

Graphical analysis of reliability data is based simply on the concept of rectifying the data in such a way that approximately straight lines can be generated when the data are plotted. Then, the graph can be quickly checked to determine if a straight line can reasonably fit the data. If not, the assumed distribution is rejected and another may be tried. The distributions we discuss in this text can all be analyzed in a graphical manner. So if one form does not fit, we can experiment with another representation.

The techniques we present here will be illustrated first for the exponential distribution. However, the procedures are general, and the later sections contain examples involving the Weibull, normal, and lognormal distributions.

We will show the techniques for the exponential distribution using several variations. Two methods for handling exact times to failure will be discussed. The third variation involves readout data. These examples will also show

how to estimate the failure rate, λ, by using the slope of the straight line drawn through the data.

Exact Failure Times

Method 1: Cumulative Hazard Method

An extensive development of this topic is given by Nelson [1972]. We will illustrate the cumulative hazard method on the times-to-failure data shown in Table 6.1. The unit of time is hours. Of the 20 components in the experiment, all eventually failed.

Recall from Chapter 2 that the cumulative hazard function H(t) of any distribution is related to the CDF, $F(t)$, by

$$H(t) = -\ln[1 - F(t)]$$

For the exponential distribution,

$$F(t) = 1 - e^{-\lambda t}.$$

Hence,

$$\begin{aligned} H(t) &= -\ln(e^{-\lambda t}), \\ &= \lambda t. \end{aligned}$$

So the cumulative hazard function of the exponential distribution varies linearly with time, and the proportionality constant is λ. Thus, if the exponential distribution holds for the data under analysis, a plot of the estimated $H(t)$ versus time should yield a linear fit with slope λ and intercept zero.

First, however, we need a method for estimating $H(t)$. At each time t, $H(t)$ is the sum of the individual hazard terms found by dividing the number of failures by the number of units surviving previously before that time t.

Table 6.1. Failure Times of Twenty Components Under Normal Operating Conditions (Time in Hours).

3.04	76.6	114.6	245.6
4.45	76.7	121.2	314.8
6.25	103.9	130.2	407.9
37.1	107.7	220.0	499.2
42.7	110.8	236.8	627.4

For exact, unique times to failure, the number of failures is always 1. Thus, a simple procedure is to order the failure times from lowest to highest, and then associate with each failure time the reverse rank starting with the initial sample size. For example, with 20 units, we have the first failure matched with 20, the second with 19, and so forth. Then for each time, the cumulative hazard function estimate is just the sum of the reciprocals of the reverse ranks, since there is one failure for each of the reverse ranks. Each reciprocal calculated this way is called a "hazard value." The calculations are illustrated in Table 6.2. The data are plotted on linear paper in Figure 6.5.

The fit to a straight line appears reasonable. The placement of the line to the data is done by eye. It is possible to apply regression analysis to the data to estimate the slope of the line, but generally an eyeball fit is adequate. From the slope of the line, we obtain an estimate of the failure rate of $\hat{\lambda} =$ 542%/K. The reciprocal of this slope estimates the MTTF, that is 185 hours.

Table 6.2. Cumulative Hazard Calculation, Individual Failure Times ($n = 20$).

FAILURE TIME	FAILURE NO.	REVERSE RANK, k	HAZARD VALUE, $100 \times (1/k)$	CUMULATIVE HAZARD
3.04	1	20	5.00	5.0
4.45	2	19	5.26	10.3
6.25	3	18	5.56	15.8
37.1	4	17	5.88	21.7
42.7	5	16	6.25	28.0
76.6	6	15	6.67	34.6
76.7	7	14	7.14	41.8
103.9	8	13	7.69	49.5
107.7	9	12	8.33	57.8
110.8	10	11	9.09	66.9
114.6	11	10	10.00	76.9
121.2	12	9	11.1	88.0
130.2	13	8	12.5	100.5
220.0	14	7	14.3	114.8
236.8	15	6	16.7	131.4
245.6	16	5	20.0	151.4
314.8	17	4	25.0	176.4
407.9	18	3	33.3	210.0
499.2	19	2	50.0	260.0
627.4	20	1	100.0	360.0

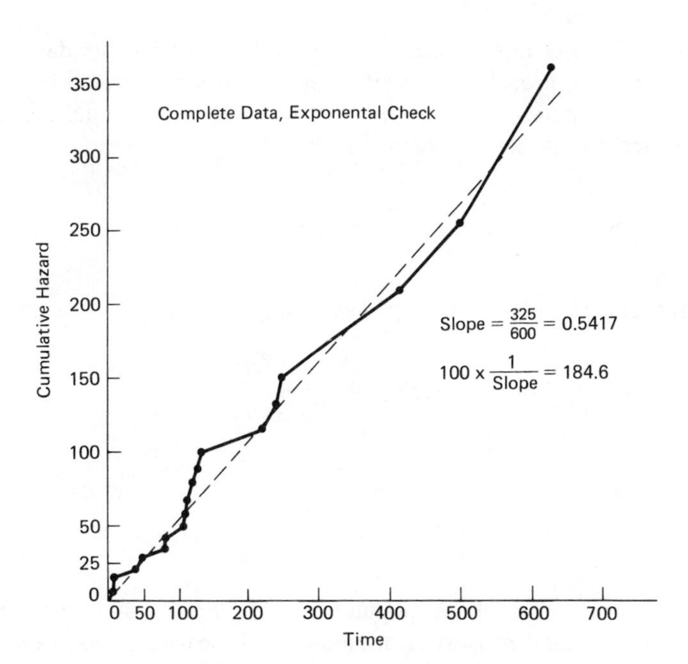

Figure 6.5. Exponential Hazard Plot, Exact Times.

Method 2: CDF Procedure

Recall that the CDF of the exponential distribution is given by

$$F(t) = 1 - e^{-\lambda t}.$$

Rewriting this equation and taking natural logarithms, we get

$$1 - F(t) = e^{-\lambda t},$$
$$-\ln[1 - F(t)] = \lambda t,$$

or

$$\ln \frac{1}{1 - F(t)} = \lambda t.$$

So if $\ln\{1/[1 - F(t)]\}$ is plotted against the time on linear by linear graph paper, or equivalently, $1/[1 - F(t)]$ is plotted versus the time on semilog (i.e., log-linear) paper, then the data should approximately fall on a straight line if the exponential distribution applies.

The main issue is now to estimate the CDF $F(t)$ from the data of exact times. A simple approach for a starting sample of size n would be to assign $1/n$ for the estimate of $F(t)$ at the first ordered failure time; $2/n$, at the second ordered time; and so forth. In general,

$$F(t_i) = \frac{i}{n} \qquad i = 1, 2, 3, \ldots, n.$$

Other authors (Hahn and Shapiro, 1967) suggest using

$$F(t_i) = \frac{i}{n+1}$$

or

$$F(t_i) = \frac{i - \frac{1}{2}}{n}$$

as preferred estimators of the population CDF $F(t)$. Our choice, because of desirable statistical properties discussed in the paper by Johnson (1951), is to use what are called "median ranks," especially when the formula can be conveniently programmed on a computer. The following equation can be derived to estimate the CDF $F(t)$;

$$F(t_i) = \frac{0.5^{1/n}(2i - n - 1) + n - 1}{n - 1} \qquad i = 1, 2, 3, \ldots.$$

Fortunately there is a simple approximation to this formula given by

$$F(t_i) = \frac{i - 0.3}{n + 0.4} \qquad i = 1, 2, 3, \ldots.$$

We recommend that the above formula be used for estimating the population CDF in all reliability plotting.

Using the same data as before, we present the necessary calculations as Table 6.3. Figure 6.6 shows the plot on semilog paper. Again a straight line is reasonable. The failure rate is estimated from the slope of the line, and 530%/K is obtained.

Note that there is an equivalence between the CDF and cumulative hazard procedures since we have shown

$$H(t) = -\ln[1 - F(t)].$$

Table 6.3. CDF Estimation by Median Ranks, Individual Failure Times ($n = 20$).

FAILURE TIME t	FAILURE COUNT i	MEDIAN RANK, $F = (i - .3)/(n + .4)$	TRANSFORMATION, $1/(1 - F)$
3.04	1	.034	1.035
4.45	2	.083	1.091
6.25	3	.132	1.152
37.1	4	.181	1.221
42.7	5	.230	1.299
76.6	6	.279	1.388
76.7	7	.328	1.489
103.9	8	.377	1.606
107.7	9	.426	1.743
110.8	10	.476	1.906
114.6	11	.524	2.103
121.2	12	.574	2.345
130.2	13	.623	2.650
220.0	14	.672	3.046
236.8	15	.721	3.580
245.6	16	.770	4.343
314.8	17	.819	5.519
407.9	18	.868	7.567
449.2	19	.917	12.03
627.4	20	.966	29.36

However, the CDF procedure is often preferred by reliability engineers since the results follow from a simple transformation of the data. Unlike the CDF, the cumulative hazard has no direct interpretation, and so it becomes more difficult for an engineer to explain his analysis to other workers. Neverthless, the cumulative hazard is a very useful mathematical concept, especially for handling multicensored data.

We plotted complete data (all units fail), but time censored data can similarly be plotted up to the last available exact times of failure. The analysis procedure is identical to that applied to complete data.

Interval Data

When data occur in the form of readout intervals, the CDF plotting method is applied more frequently than the cumulative hazard procedure because of the more direct interpretation of the CDF. However, the concept of median ranks, used to provide accurate plotting positions when exact failure times

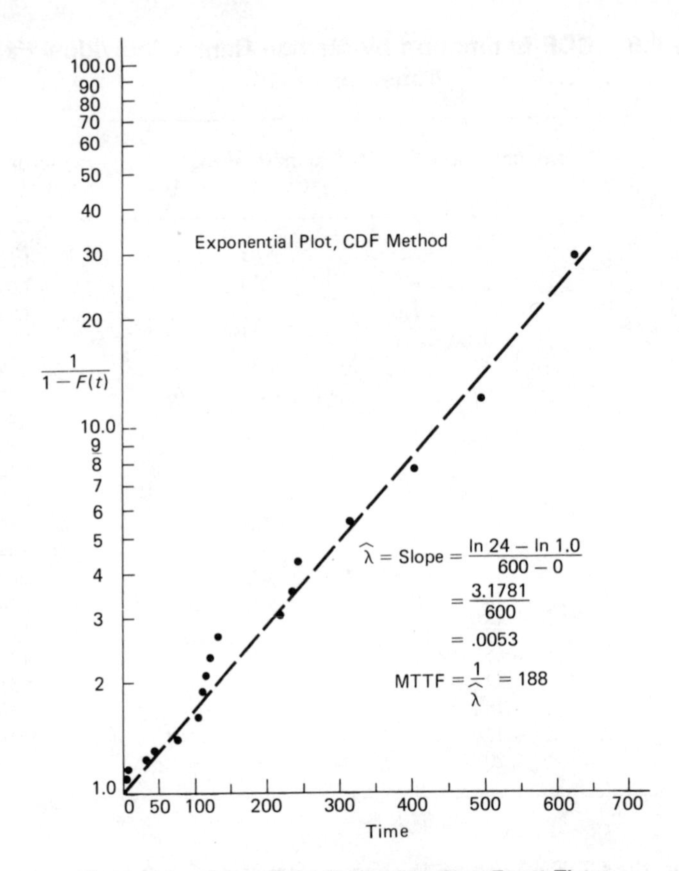

Figure 6.6. Exponential Probability Plot, Exact Times.

are known is no longer needed. The best CDF estimate at a given readout time is just the cumulative number of failures up to that time divided by the starting sample size.

We show in Table 6.4 some readout times for a group of 100 components stressed in a study that ran to 600 hr. Here we have censored data with two components still surviving at the end of the experiment. The calculations are also shown in Table 6.4. Note the CDF is estimated by

$$F(t) = \frac{\text{total failures by time } t}{\text{starting sample size}}.$$

The graph is shown as Figure 6.7. The failure rate is estimated again by the slope as 618%/K.

Table 6.4. Readout Data (n = 100).

INTERVAL, HR	FREQUENCY OF FAILURES	CUMULATIVE FREQUENCY	ESTIMATED CDF, $F(t)$	TRANSFORMATION, $1/(1-F)$
0–50	29	29	.29	1.409
50–100	21	50	.50	2.000
100–150	11	61	.61	2.564
150–200	10	71	.71	3.448
200–250	4	75	.75	4.000
250–300	8	83	.83	5.882
300–350	6	89	.89	9.091
350–400	3	92	.92	12.50
400–450	1	93	.93	14.29
450–500	3	96	.96	25.00
500–550	1	97	.97	33.33
550–600	1	98	.98	50.00

Alternative Estimate of the Failure Rate and Mean Life

We note that at $t = 1/\lambda$, $F(1/\lambda) = 1 - e^{-1} = .63212...$ So an alternative estimator of the MTTF $= 1/\lambda$ can be found by noting the time where the plotted line intersects the horizontal line obtained at $F(t) = .63212...$, or equivalently where the transformed variable $1/[1 - F(\text{MTTF})] = e = 2.7183...$ Such a procedure is a quick way to get the MTTF without calculating the slope. It may be difficult to read off the MTTF value accurately using this procedure, but in many cases it will suffice. Here, we estimate from Figure 6.7 that the MTTF = 160 hr or $\lambda = 625\%/\text{K}$.

PROBABILITY PLOTTING FOR THE WEIBULL DISTRIBUTION

The equation for the Weibull distribution is

$$F(t) = 1 - e^{-(t/c)^m},$$

where c, m, and $t \geq 0$. As was done in Chapter 4, we rewrite this equation in the form

$$1 - F(t) = e^{-(t/c)^m}$$

and take natural logarithms of both sides twice to get

$$\ln\{-\ln[1 - F(t)]\} = m \ln t - m \ln c.$$

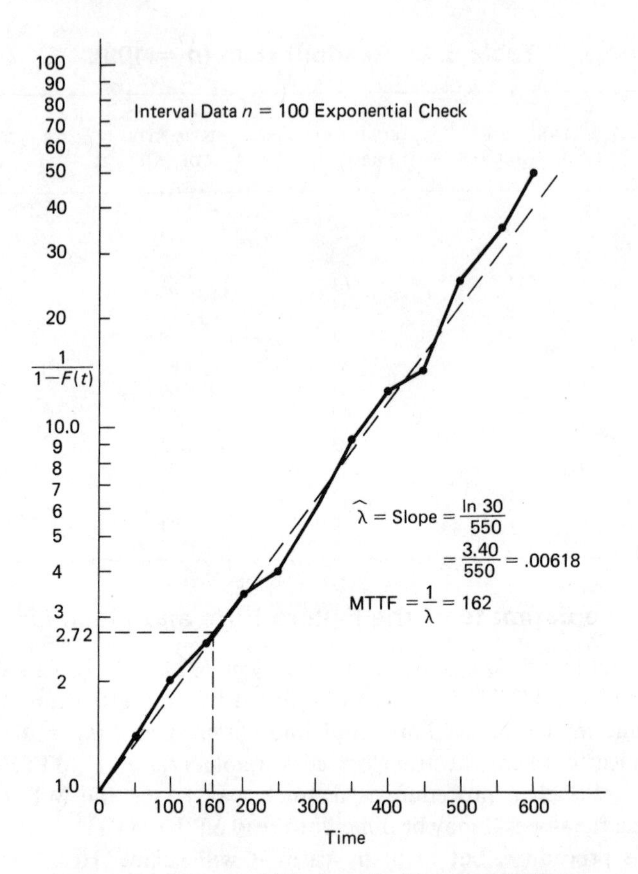

Figure 6.7 Exponential Probability Plot, Interval Data.

If we now rename the variable $y = \ln\{-\ln[1 - F(t)]\}$ and $x = \ln t$, then it is obvious the above transformed equation represents a straight line with slope m and intercept $-m \ln c$. Now we use the fact from Chapter 2 that the cumulative hazard $H(t)$ is related to the CDF $F(t)$ by the expression

$$H(t) = -\ln[1 - F(t)].$$

So the transformed Weibull equation can be written in terms of the cumulative hazard as

$$\ln H(t) = \text{m} \ln t - m \ln c.$$

Thus, a plot of the log of the cumulative hazard versus the log of time should approximate a straight line with slope m and intercept $m \ln c$ if the assumed Weibull model applies.

The previous discussion shows that several options are available for analyzing Weibull data graphically:

1. Plot $H(t)$ versus t on log–log paper, using the reverse rank procedure for estimating $H(t)$.
2. Plot $-\ln[1 - F(t)]$ versus t on log–log paper, using the median ranks to estimate $F(t)$.

In addition, special-purpose Weibull graph paper is available that permits direct plotting of the estimated CDF versus time. These papers also have attached scales that allow quick parameter estimation. Since the first two methods are basically equivalent, we shall illustrate Weibull plotting using the second procedure, and then show how to use a special purpose paper on the same data.

Weibull Plotting—Exact Failure Times—CDF Method

EXAMPLE 6.4 WEIBULL CDF PLOTTING—EXACT TIMES

Twenty test vehicles to investigate dielectric breakdown strength are placed under stress. The test is concluded after 600 hr at which time 18 units have failed. The data are shown in Table 6.5. Plot the results to determine if the Weibull distribution is a reasonable fit to the data. Estimate from the graph the parameters of the distribution: shape factor m and characteristic life c.

SOLUTION

For exact, unique failures times, the estimation of $F(t)$ is done by the same procedure used in the exponential case, that is via the median ranks as shown in Table 6.3. The Weibull plot is shown as Figure 6.8. The fit appears reasonable to the model.

To estimate the slope from the log–log paper plot, we first pick two points on the log time axis, say $t_1 = .1$ and $t_2 = 1000$ hr, and find the corresponding $Y_1 = -\ln[1 - F(t)]$ on the vertical log axis to be $Y_1 = .03$ and $Y_2 = 3.2$. The slope is then evaluated from the equation

$$m = \frac{\ln(Y_2/Y_1)}{\ln(t_2/t_1)} = \frac{\ln(3.2/.03)}{\ln(1000/.1)} \approx .507.$$

The intercept $-m \ln c$ is found at $t = 1$ (where $\ln 1 = 0$, thus giving the intercept) to be .098; hence, $c = (.098)^{-1/.507} \approx 98$.

Table 6.5. Weibull Example (n = 20).

FAILURE TIME, t	FAILURE COUNT, i	MEDIAN RANK, $F = (i - .3)/(n + .4)$	TRANSFORMATION, $-\ln(1 - F)$
.69	1	.034	.035
.94	2	.083	.087
1.12	3	.132	.142
6.79	4	.181	.200
9.28	5	.230	.262
9.31	6	.279	.328
9.95	7	.328	.398
12.90	8	.377	.474
12.93	9	.426	.556
21.33	10	.475	.645
64.56	11	.525	.743
69.66	12	.574	.852
108.38	13	.623	.974
124.88	14	.672	1.114
157.02	15	.721	1.275
190.19	16	.770	1.469
250.55	17	.819	1.708
552.87	18	.868	2.024

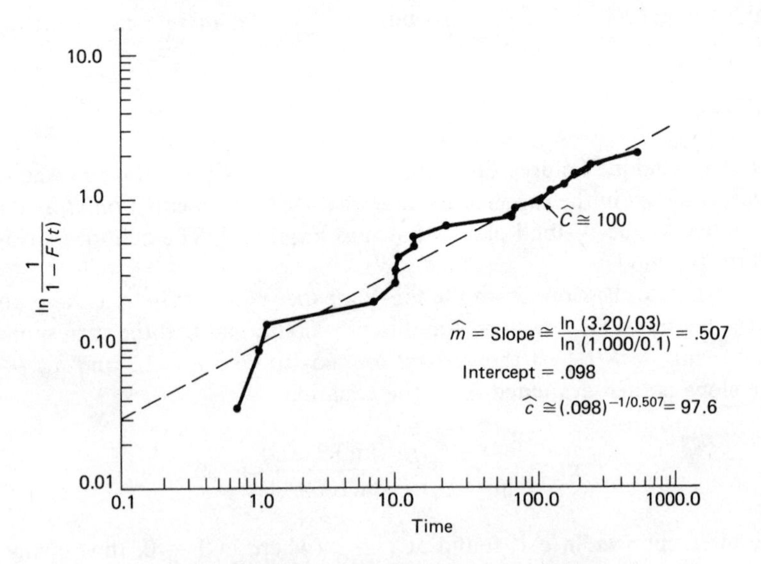

Figure 6.8. Weibull Probability Plot, Exact Times.

Note an alternative quick estimator of the characteristic life c is found at the CDF value for which $F(c) = 1 - e^{-1} = .63212 \ldots$ or $-\ln[1 - F(c)] = 1.00$. From Figure 6.8, we get $c \approx 100$.

Instead of plotting the data on log–log paper, special-purpose charts are available (TEAM, Tamworth, NH is one source) to facilitate the Weibull analysis. Using the same data in Table 6.5, we plot the CDF median rank estimate versus time on one type of Weibull probability paper in Figure 6.9. A straight line has been visually drawn through the points. The fit appears acceptable. To estimate the shape factor m, we now draw a line, starting at the small circle marked origin in the upper left-hand corner of the graph, parallel to the line through the data points. The intersection of that line with the scale on the left gives a direct estimate of m. Here, $m \approx .52$. The characteristic life c is estimated by finding the intersection of the line through the points with the line at the percent failure level of 63.2, marked as a small circle in the figure. Here, $c \approx 96$ hr.

We used one form of Weibull paper. Other types involving direct plotting of the estimated cumulative hazard values are also available. However, we emphasize that for all distributions we treat in this text, one can make probabil-

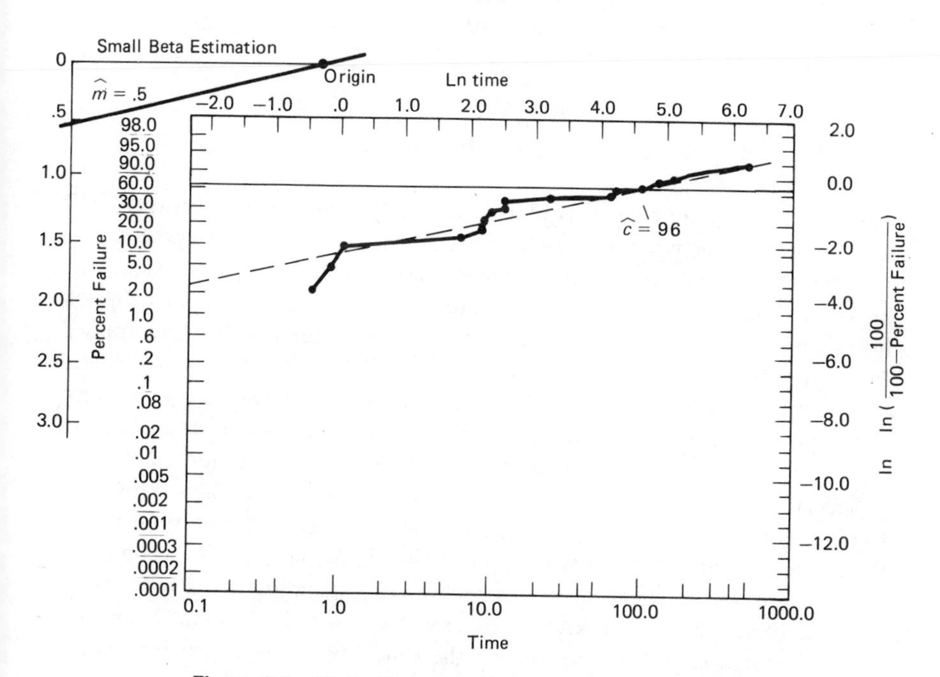

Figure 6.9. Weibull Plot on Special Chart, Exact Times.

ity plots from common linear or logarithmic papers by applying the transformations we show and possibly using some common tables.

PROBABILITY PLOTTING FOR THE NORMAL AND LOGNORMAL DISTRIBUTIONS

Normal Distribution

Let z be the standard normal variate, that is, $z = (x - \mu)/\sigma$, where (μ,σ^2) are the population mean and variance of a normal distribution. Recall that there is a one-to-one correspondence between z and the area under the normal PDF curve. For example, $z = 0$ corresponds to 50%; $z = -1.28$ to 10%; and so on. If the failure times of a population are normally distributed, then we can associate a z value with a given percent failure, that is, the percent failure will uniquely correspond to a z value. Following the notation in Chapter 5, designate this relation by the equation $z = \Phi^{-1}(F)$, with $\Phi^{-1}(F)$ denoting the inverse CDF function. Thus, we can write

$$\Phi^{-1}(F) = \frac{x - \mu}{\sigma},$$

or

$$x = \sigma\Phi^{-1}(F) + \mu.$$

This last relation shows if x is plotted on linear graph paper versus $\Phi^{-1}(F)$ for the normal distribution, one will get a straight line with slope equal to the standard deviation σ and intercept (at $z = 0$ or $F = .5$) equal to the mean μ. Thus, as we showed in Example 5.3, if the normal distribution holds, sample data of failure times plotted against the z value [determined from a normal table by treating the estimated CDF $F(x)$ as an area] should approximate a straight line. The slope of the line estimates σ; the intercept (50% point) estimates the mean μ. This method allows us to obtain an estimate of the mean and standard deviation from censored normal data.

Special-purpose normal probability plotting papers are available. However, using a normal table, one can generate the z values and plot time versus the z values on linear by linear paper. In either case, the slope can be obtained between any two convenient points whose corresponding areas represent a multiple of z. That is, we know that area for $z = -1.00$ is .1587, and for $z = 0$, the area is .5. So if we find on the time axis the times corresponding to 15.87% and 50%, the difference in these times is an estimate of the standard deviation σ. Similarly we could take the difference between the times at percent failures of 2.28% ($z = -2.00$) and 97.72% ($z = 2.00$) and divide the difference by 4 to get an estimate of the standard deviation.

If one has highly censored data, and must use a low percent of failures to estimate the slope, the mean can be estimated by substituting the estimated slope σ into the equation $\mu = \sigma\Phi^{-1}(F) + t$ for any convenient time and percent failure pair. Thus, extrapolation to the 50% failure point to estimate the mean is unnecessary.

Lognormal Distribution

As discussed in Chapter 5, if the natural logarithms of failure times are normally distributed with mean μ and variance σ^2, then the distribution of the times is lognormal with median $T_{50} = e^{\mu}$ and shape parameter σ. The median T_{50} and shape parameter σ are the standard parameters for expressing the lognormal distribution.

Similar to the normal, the rectifying equation is found by using the standard normal variate in the form

$$z = \frac{\ln t - \ln T_{50}}{\sigma}$$

or since $z = \Phi^{-1}(F)$,

$$\ln t = \sigma\Phi^{-1}(F) + \ln T_{50}.$$

Thus, if the lognormal distribution applies, a plot of the natural logarithms of times versus the inverse transformation of the estimated CDF percent failures (that is, the z value corresponding to a given area) should approximate a straight line which can be used to estimate the shape parameter σ from the slope, and the T_{50} from the fiftieth percentile point.

Again we have many options. Using the term "percent failures" loosely to refer to the estimated CDF values, we can plot: (a) corresponding to the percent failures, the z values obtained from a normal table versus the log of times on linear–linear paper; (b) the percent failures versus the log of times on normal probability paper; (c) the percent failures versus the times on lognormal probability paper; or (d) the estimated cumulative hazard versus the times on lognormal hazard paper. All methods are basically equivalent, but computationally, the use of either lognormal probability or hazard paper is easiest.

The slope can be estimated from the equation

$$\sigma = \frac{\ln(t_2/t_1)}{\Phi^{-1}[F(t_2)] - \Phi^{-1}[F(t_1)]},$$

or since $z = \Phi^{-1}(F) = 0$ at $F = 50\%$ and $z = -1.00$ at $F = 15.9\%$,

$$\sigma = \ln\frac{T_{50}}{T_{15.9}}.$$

As in the normal case, any convenient z multiple can also be used to estimate σ. Similarly, if T_{50} is outside of the range of the graph, the slope can be estimated and T_{50} solved from the equation

$$T_{50} = te^{-\sigma\Phi^{-1}[F(t)]} = te^{-\sigma z}$$

using any convenient time and percent failure combination.

EXAMPLE 6.5 LOGNORMAL CDF PLOTTING

Six hundred transistors are tested under stress for 1000 hr. Each failure is individually recorded in time. At the end of the experiment, 17 units have

Table 6.6. Lognormal Example (n = 600).

FAILURE TIME, t	FAILURE COUNT, i	MEDIAN RANK $F = (i - .3)/(n + .4)$
3.7	1	.0012
25.9	2	.0028
58.6	3	.0045
78.4	4	.0062
146.7	5	.0078
162.3	6	.0095
224.1	7	.0112
228.6	8	.0128
275.9	9	.0145
282.9	10	.0162
481.4	11	.0178
689.5	12	.0195
720.0	13	.0215
770.0	14	.0228
851.2	15	.0245
871.7	16	.0261
999.3	17	.0278

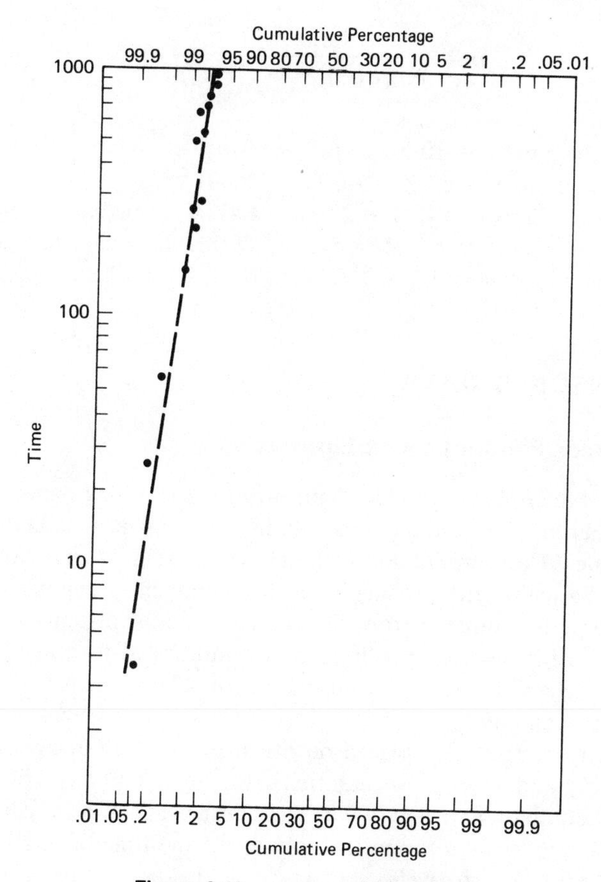

Figure 6.10. Lognormal Example.

failed for a particular failure mode. The data are given in Table 6.6. It is assumed that the lognormal distribution applies. Check this assumption by plotting the data on lognormal probability paper. Estimate the parameters of the distribution. How long a test would be required to get 10% of the product to fail?

SOLUTION

The estimated CDF at each point using median ranks is shown in Table 6.6. The data are graphed as Figure 6.10. The slope is estimated from the equation

$$\sigma = \frac{\ln(t_2/t_1)}{\Phi^{-1}[F(t_2)] - \Phi^{-1}[F(t_1)]},$$

by choosing the pairs $t = 10$ hr, $F(10) = .00165$, implying $\Phi^{-1}(F) = -2.95$ from the normal table, and $t = 1000$ hr, $F(1000) = .0275$, $\Phi^{-1}(F) = -1.92$. Then, $\sigma = \ln(1000/10)/(-1.92 + 2.95) \approx 4.47$. The median is found from $T_{50} = te^{-\sigma z} = 1000e^{-4.47 \times -1.92} \approx 5.3 \times 10^6$ hr. The time to reach 10% failures is $t = T_{50}e^{\sigma z_{.10}} = 5.3 \times 10^6 e^{4.47 \times -1.28} \approx 17{,}350$ hr.

MULTICENSORED DATA

Kaplan–Meier Product Limit Estimation

The Kaplan–Meier (1958) product limit estimator is a long name for a very simple and useful procedure for calculating survival probability estimates at various times. The power of this method lies in its ability to handle multicensored data. Basically, the probability of a component surviving during an interval of time is estimated from the observed fallout of units starting the interval. The estimates for each interval are multiplied together to provide a survival or reliability estimate from time zero. The procedure is best illustrated with an example.

Suppose eight objects are placed on life stress and failures occur at 200, 300, 400, 500, and 600 hr. In addition, two good units are removed for destructive examination: one at 200 hr and another at 400 hr. Such nonfailed units effectively provide information only up to the time of removal, called a censoring time. Alternatively, we state that "losses" occurred at 200 and 400 hr. Another common terminology is to state that these units have "running times" of 200 and 400 hr.

Since we started with 8 units and 1 failed by 200 hr, the probability estimate for a unit surviving 200 hr is $7/8 = .875$. At 200 hr, we also had a loss. Thus, 6 units remained on stress until the next failure at 300 hr. The probability of surviving from 200 to 300 hr is estimated by $5/6$. The probability of surviving from time zero to 300 hr is estimated by

$$P(T > 300) = (7/8)(5/6) = .729.$$

In words, the probability of surviving to 500 hr is given by the probability of surviving to 200 hr times the probability of surviving to 300 hr given survival to 200 hr.

Similarly, the survival probabilities at the failure times are estimated below:

TIME, hr	SURVIVAL ESTIMATE (RELIABILITY)
400	$(7/8)(5/6)(4/5) = .583$
500	$(7/8)(5/6)(4/5)(2/3) = .389$
600	$(7/8)(5/6)(4/5)(2/3)(1/2) = .194$
700	$(7/8)(5/6)(4/5)(2/3)(1/2)(0/1) = .000$

The cumulative distribution function can be estimated at each failure time by subtracting the survival values from 1.000. Thus, from knowledge of the failure and censoring times we can construct an estimate of the CDF for multicensored data. Then, the estimated CDFs at each failure time can be plotted on any type probability paper to test the applicability of any distributional form, using the methods described earlier in this chapter.

Cumulative Hazard Estimation

An alternative procedure for handling multicensored data using the cumulative hazard function is described by Nelson (1983). This technique involves

Table 6.7. Cumulative Hazard Calculation.

TIME OF FAILURE OR READOUT	NUMBER ON TEST JUST PRIOR TO FAILURE(S)	NUMBER OF NEW FAILURES	HAZARD VALUE	CUMULATIVE HAZARD VALUE
T_1	N_1	r_1	$h_1 = \frac{r_1}{N_1}$	$H_1 = 100 \times h_1$
T_2	N_2	r_2	$h_2 = \frac{r_2}{N_2}$	$H_2 = H_1 + 100 \times h_2$
T_3	N_3	r_3	$h_3 = \frac{r_3}{N_3}$	$H_3 = H_2 + 100 \times h_3$
.	.	.	.	.
.	.	.	.	.
.	.	.	.	.
T_k	N_k	r_k	$h_k = \frac{r_k}{N_k}$	$H_k = H_{k-1} + 100 \times h_k$

estimating the cumulative hazard function at each failure time and then plotting the points on cumulative hazard paper. The method of estimation is detailed in Table 6.7.

EXAMPLE 6.6 CUMULATIVE HAZARD PLOTTING

Using the same data as in the Kaplan–Meier section, calculate the cumulative hazard estimates at each failure time.

SOLUTION

FAILURE TIMES	NUMBER ON TEST	NUMBER OF FAILURES	HAZARD VALUE	CUMULATIVE HAZARD VALUE
200	8	1	$\frac{1}{8} = .125$	12.5
300	6	1	$\frac{1}{6} = .167$	29.2
400	5	1	$\frac{1}{5} = .200$	49.2
500	3	1	$\frac{1}{3} = .333$	82.5
600	2	1	$\frac{1}{2} = .500$	132.5

The above cumulative hazard values could now be plotted on hazard paper to check against a specific distribution type.

SUMMARY

This chapter has reviewed graphical procedures for supplementing the analytical tools covered in previous chapters. We have discussed straight line concepts, least-squares regression, and linear rectification. We have shown how various distribution models could be checked for applicability by the use of probability or hazard plotting. In addition, we have demonstrated how quick parameter estimates could be obtained from the graphs. Special commercial probability papers can be used or equivalent results may be obtained with linear or log papers coupled with simple transformations of CDF estimates. These methods are easily applied to censored data. Also, with Kaplan–Meier or cumulative hazard procedures, we can analyze multicensored data.

7

Physical Acceleration Models

If we have enough test data, the methods described in the preceding chapters will allow us to fit our choice of a life distribution model and estimate the unknown parameters. However, with today's highly reliable components, we are often unable to obtain a reasonable amount of test data when stresses approximate normal use conditions. Instead, we force components to fail by testing at much higher than the intended application conditions. By this way, we get failure data that can be fitted to life distribution models, with relatively small test sample sizes and practical test times.

The price we have to pay for overcoming the dilemma of not being able to estimate failure rates by testing directly at use conditions (with realistic sample sizes and test times) is the need for additional modeling. How can we go from the failure rate at high stress to what a future user of the product is likely to experience at much lower stresses?

The models used to bridge the stress gap are known as acceleration models. This chapter develops the general theory of these models and looks in detail at what acceleration testing under the exponential or Weibull or lognormal means. Several well-known forms of acceleration models, such as the Arrhenius and the Eyring, are described. Practical use of these models, both for actual failure data and for degradation data, is discussed.

ACCELERATED TESTING THEORY

The basic concept of acceleration is simple. We hypothesize that a component, operating under the right levels of increased stress, will have exactly the same failure mechanisms as seen when used at normal stress. The only difference is "things happen faster." For example, if corrosion failures occur at typical use temperatures and humidities, then the same type of corrosion happens much quicker in a humid laboratory oven at elevated temperature.

In other words, we can think of time as "being accelerated," just as if the process of failing were filmed and then played back at a faster speed. Every step in the sequence of chemical or physical events leading to the failure state occurs exactly as at lower stresses; only the time scale measuring event duration has been changed.

When we find a range of stress values over which this assumption holds, we say we have *true acceleration.* From the film replay analogy, it is clear that true acceleration is just a transformation of the time scale. Therefore, if we know the life distribution for units operating at a high laboratory stress, and we know the appropriate time scale transformation to a lower stress condition, we can mathematically derive the life distribution (and failure rate) at that lower stress. This is the way we will proceed.

In theory, any well-behaved (order preserving, continuous, etc.) transformation could be a model for true acceleration. However, in terms of practical applicability, we almost always restrict ourselves to simple constant multipliers of the time scale. *When every time of failure and every distribution percentile is multiplied by the same constant value to obtain the projected results at another operating stress, we have linear acceleration.*

Under a linear acceleration assumption, we have the relationship (time to fail at stress S_1) = AF × (time to fail at stress S_2), where AF is the acceleration constant relating times to fail at the two stresses. "AF" is called the acceleration factor between the stresses.

If we use subscripts to denote stress levels, with U being a typical use set of stresses and S (or S_1, S_2, . . .) for higher laboratory stresses, then the key equations in Table 7.1 hold no matter what the underlying life distribution happens to be.

In Table 7.1, t_U represents a random time to fail at use conditions, while t_S is the time the same failure would have happened at a higher stress. Similarly, F_U, f_U, and h_U are the CDF, PDF, and failure rate at use conditions, while F_S, f_S, and h_S are the corresponding functions at stress S.

Table 7.1. General Linear Acceleration Relationships.

1. Time to fail:	$t_U = \mathrm{AF} \times t_S$
2. Failure probability:	$F_U(t) = F_S\left(\frac{t}{\mathrm{AF}}\right)$
3. Density function:	$f_U(t) = \left(\frac{1}{\mathrm{AF}}\right) f_S\left(\frac{t}{\mathrm{AF}}\right)$
4. Failure rate:	$h_U(t) = \left(\frac{1}{\mathrm{AF}}\right) h_S\left(\frac{t}{\mathrm{AF}}\right)$

Equation 1 of Table 7.1 is the linear acceleration assumption and the other equations all follow directly from this and the definitions of the functions involved. For example, $F_U(t)$ is the probability of failing by time t at use stress, which is equivalent to failing at time (t/AF) at stress S (by equation 1). This relationship is stated in equation 2 and the next two equations follow by standard change of variable methods. (For example, see Mendenhall and Scheaffer, 1973.)

Table 7.1 gives the mathematical rules for relating CDFs and failure rates from one stress to another. These rules are completely general, and depend only on the assumption of true acceleration and linear acceleration factors. In the next three sections, we will see what happens when we apply these rules to exponential or Weibull or lognormal life distributions.

EXPONENTIAL DISTRIBUTION ACCELERATION

We add the assumption that $F_S(t) = 1 - e^{-\lambda_s t}$. In other words, times to fail at high laboratory stresses can be modeled by an exponential life distribution with failure rate parameter λ_s. Using equation 2 to derive the CDF at use condition, we get

$$F_U(t) = F_S\left(\frac{t}{\mathrm{AF}}\right) = 1 - e^{-\lambda_s t/\mathrm{AF}} = 1 - e^{-(\lambda_s/\mathrm{AF})t}.$$

By letting $\lambda_u = \lambda_s/\mathrm{AF}$, we see that the CDF at use conditions remains exponential, with new parameter λ_s/AF.

This equation demonstrates that an exponential fit at any one stress condition implies an exponential fit at any other stress within the range where true linear acceleration holds. Moreover, when time is multiplied by an acceleration factor AF, the failure rate is reduced by dividing by AF.

The fact that the failure rate varies inversely with the acceleration factor sometimes misleads engineers to assume this is always the case with linear acceleration. This is not correct. In general, the failure rate changes in a very nonlinear fashion under linear acceleration of the time scale. The simple results of this section apply only for the exponential distribution.

EXAMPLE 7.1 EXPONENTIAL ACCELERATION FACTORS

A component, tested at 125°C in a laboratory, has an exponential distribution with MTTF 4500 hr. Normal use temperature for the component is 25°C. Assuming an acceleration factor of 35 between these two temperatures, what will the use failure rate be and what percent of these components will fail before the end of the expected useful life period of 40,000 hr?

SOLUTION

The MTTF is the reciprocal of the failure rate and varies directly with the acceleration factor. Therefore the MTTF at 25°C is $4500 \times 35 = 157{,}500$. The use failure rate is $1/157{,}500 = .635\%/\mathrm{K}$. The cumulative percent of failures at 40,000 hr is given by $1 - e^{-.00635 \times 40} = 22.4\%$.

WEIBULL DISTRIBUTION ACCELERATION

This time we start with the assumption that $F_S(t)$ follows a Weibull distribution with characteristic life c_S and shape parameter m_S. The equation for the CDF is

$$F_S(t) = 1 - e^{-(t/c_S)^{m_S}}$$

and, transforming to use stress, we have

$$F_U(t) = F_S\left(\frac{t}{\mathrm{AF}}\right) = 1 - e^{-[(t/\mathrm{AF})/c_S]^{m_S}}$$

$$= 1 - e^{-[t/(\mathrm{AF} \times c_S)]^{m_S}} = 1 - e^{-(t/c_U)^{m_U}},$$

where $c_U = \mathrm{AF} \times c_S$ and $m_U = m_S = m$.

This result shows that if the life distribution at one stress is Weibull, the life distribution at any other stress (assuming true linear acceleration) is also Weibull. The shape parameter remains the same while the characteristic life parameter is multiplied by the acceleration factor.

The equal shape result is highly significant. It is often mistakenly added as an assumption, in addition to assuming a linear acceleration model and a Weibull life distribution. As we have seen, however, it is a necessary mathematical consequence to the other two assumptions. If different stress cells yield data with very different shape parameters, then either we do not have true linear acceleration or the Weibull distribution is the wrong model for the data.

In Chapter 6, when we discussed Weibull data plotting we saw that the shape parameter turned out to be the slope of the line fitted to the data, when plotted on an appropriate type of graph paper. Therefore, when we have Weibull acceleration and we plot several stress cells of data on the same sheet of graph paper, we should end up with parallel lines. The lines will not be exactly parallel, of course, since we are dealing with sample data and estimates of the underlying distributions. However, lines that are very far from appearing parallel would indicate either model or data problems.

By calculating the Weibull failure rate at stress and use conditions, it is easy to see how it varies under acceleration. For a stress failure rate we have $h_s(t) = (m/c_s)(t/c_s)^{m-1}$. By expressing the characteristic life parameter for use as $c_U = \mathrm{AF} \times c_S$, we obtain

$$h_U(t) = \frac{m}{\mathrm{AF}\cdot c_S}\left(\frac{t}{\mathrm{AF}\cdot c_S}\right)^{m-1} = \frac{1}{\mathrm{AF}^m}\frac{m}{c_S}\left(\frac{t}{c_S}\right)^{m-1} = \frac{h_S(t)}{\mathrm{AF}^m}.$$

This is a linear change in the failure rate, but the multiple is $1/\mathrm{AF}$ only when $m = 1$ and the distribution is exponential; otherwise the failure rate is multiplied by $1/\mathrm{AF}^m$.

EXAMPLE 7.2 WEIBULL MULTIPLE STRESS CELLS

Random samples of a manufacturers capacitors were tested to determine how temperature accelerates failure times. Three temperatures were used; one cell at 85°C, one at 105°C, and one at 125°C. Each cell, or controlled oven, contained 40 capacitors operating at the oven temperature. At the following readout times new failures were determined: 24, 72, 168, 300, 500, 750, 1000, 1250, and 1500 hr. All testing was completed in about 10 weeks, with results as given in Table 7.2.

Assuming a Weibull distribution applies, plot all the three cells of test data on the same sheet of Weibull paper, estimate parameters, and check whether the equal slope (same m) consequence of true acceleration looks reasonable. If maximum likelihood data analysis programs are available, also use these to estimate parameters and test for equal shapes. Use the cell charac-

Table 7.2. Weibull Temperature Stress Failure Data.

READOUT TIME	85°C NEW FAILURES	105°C NEW FAILURES	125°C NEW FAILURES
24	1	2	5
72	0	1	10
168	0	3	13
300	1	2	2
500	0	2	3
750	3	4	2
1000	0	5	2
1250	1	1	1
1500	2	4	0
Total	8	24	38

teristic life estimates to compute acceleration factors between 85 and 105°C, 85 and 125°C, and 105 and 125°C.

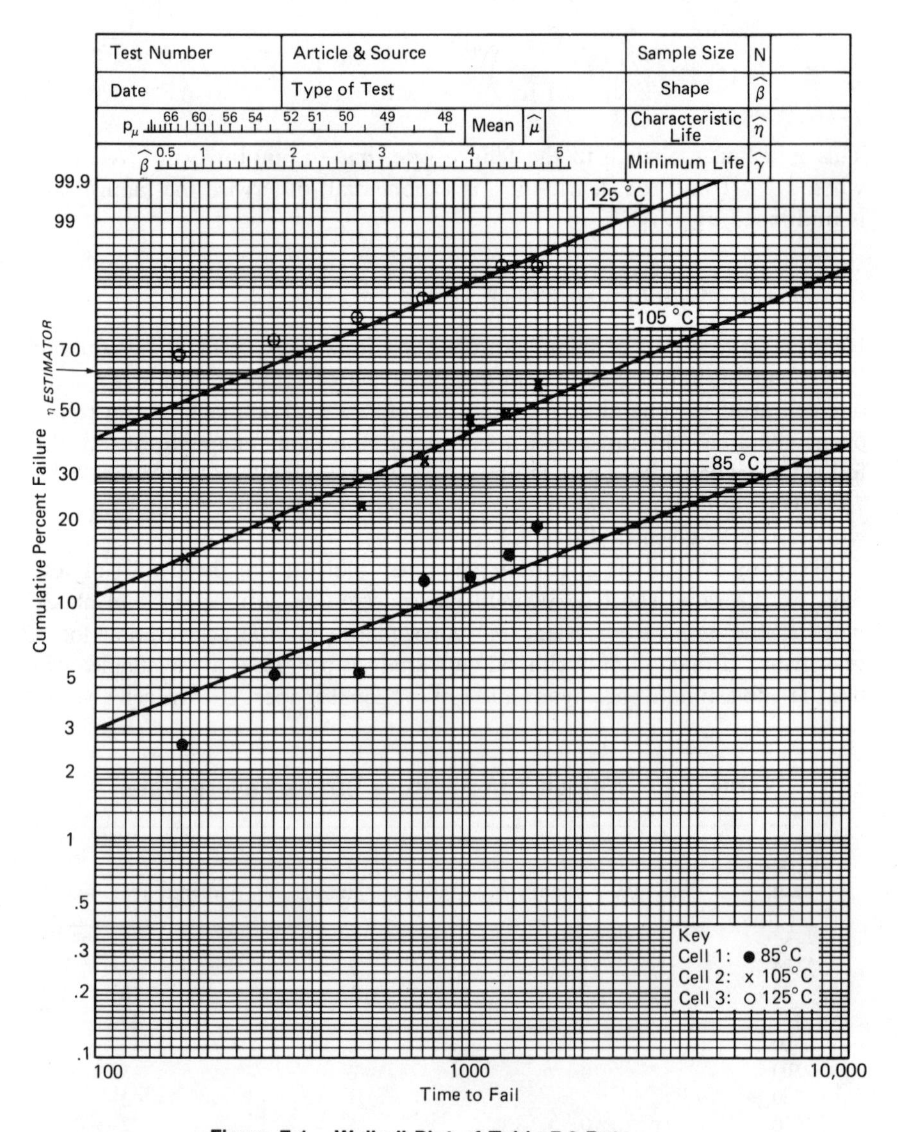

Figure 7.1. Weibull Plot of Table 7.2 Data.

SOLUTION

Figure 7.1 shows the cumulative failures in each cell plotted at the corresponding readout time. The lines fitted through the points for each cell were obtained using least squares (regression) on the equation

$$\ln\{-\ln[1-\hat{F}(t)]\} = -m \ln c + m \ln t.$$

When using a regression routine on this equation, the left-hand side, evaluated at each readout time with $\hat{F}(t)$ estimated by the cumulative fraction failures to time t, is the dependent variable. The natural logarithm of each readout time is the independent variable.

The graphical estimates are: 85°C cell—$\hat{m} = .57$, $\hat{c} = 40194$; 105°C cell—$\hat{m} = .70$, $\hat{c} = 2208$; 125°C cell—$\hat{m} = .71$, $\hat{c} = 242$.

Maximum likelihood estimates for the same cells are:

85°C cell	$\hat{m} = .81$	$\hat{c} = 9775$	acceleration to 105°C = 5.5,
105°C cell	$\hat{m} = .82$	$\hat{c} = 1746$	acceleration to 125°C = 8.0,
125°C cell	$\hat{m} = .68$	$\hat{c} = 221$	acceleration from 85°C = 44.2.

If constrained MLE is done, holding all cells to have the same shape parameter, the common $\hat{m}$ is .72, and the $\hat{c}$ estimates are 12,270, 1849, and 229 for the 85, 105, and 125°C cells, respectively.

From the plot, or the more sophisticated maximum likelihood analysis, it can be seen that the equal slope "true acceleration" assumption is reasonable.

LOGNORMAL DISTRIBUTION ACCELERATION

Now we assume that at laboratory stress $F_S(t)$ can be adequately modeled by the lognormal distribution

$$F_S(t) = \Phi\left[\frac{\ln(t/T_{50_S})}{\sigma_S}\right]$$

where Φ is the standard normal CDF defined in Chapter 5 and T_{50s} and σ_s are the lognormal parameters for the laboratory stress life distribution.

Making the acceleration transformation of the time scale given by $F_U(t) = F_S(t/\mathrm{AF})$, we find

$$F_U(t) = F_S\left(\frac{t}{\mathrm{AF}}\right) = \Phi\left\{\frac{\ln[(t/\mathrm{AF})/T_{50s}]}{\sigma_s}\right\} = \Phi\left\{\frac{\ln[t/(\mathrm{AF}\cdot T_{50s})]}{\sigma_s}\right\}$$

which is again a lognormal distribution with $\sigma_u = \sigma_s = \sigma$, and $T_{50u} = \mathrm{AF} \times T_{50s}$, where σ_u and T_{50u} are the use stress lognormal parameters.

This result is similar to that obtained for the Weibull: true linear acceleration does not change the type of distribution or the shape parameter. Only the scale parameter is multiplied by the acceleration factor between the two stresses.

Since shape or σ is equivalent to the slope on standard lognormal paper, again we expect different stress cells of data to give rise to nearly parallel lines when plotted on the same graph paper. The ratio of the times to reach any percentile, such as T_{50}, gives an estimate of the acceleration factor between the stresses.

Again it is true that the parallel lines or equal σ's result is a consequence of true linear acceleration and a lognormal life distribution model, and not an additional assumption. Also, the relationship between failure rates before and after acceleration is complicated and depends on the particular time point under evaluation. Failure rates must be calculated using the basic definition for $h(t)$ given in Chapter 5.

EXAMPLE 7.3 LOGNORMAL MULTIPLE STRESS CELLS

A semiconductor module has been observed to fail due to metal ions migrating between conductor lines and eventually causing a short. Both temperature and voltage affect the time it takes for such failures to develop. It is decided to model the kinetics of failure by conducting a stress matrix experiment using six different combinations of temperature and voltage. The stress cells are: 125°C, 8 V; 125°C, 12 V; 105°C, 12 V; 105°C, 16 V; 85°C, 12 V, 85°C, 16 V. Normal use stress is 25°C and 4 V. Sample size is 50 per cell.

Each cell is read out for new failures at 24, 100, 150, 250, 500, 750, and 1000 hr. The failure data are summarized in Table 7.3.

Table 7.3. Lognormal Stress Failure Data.

READOUT	CELL 1 125°C 8 V FAILURES	CELL 2 125°C 12 V FAILURES	CELL 3 105°C 12 V FAILURES	CELL 4 105°C 16 V FAILURES	CELL 5 85°C 12 V FAILURES	CELL 6 85°C 16 V FAILURES
24	0	0	0	0	0	0
100	1	4	0	0	0	0
150	6	3	1	1	0	0
250	6	23	3	10	1	1
500	23	13	19	21	1	5
750	9	5	15	10	3	15
1000	3	1	6	5	5	6
Total	48	49	44	47	10	27

Assuming a lognormal failure distribution, plot the six cells of data and check whether the equal shape hypothesis of "true acceleration" is reasonable. Estimate stress cell parameters by graphical methods and the method of maximum likelihood.

SOLUTION

Cumulative failure data points from the six stress cells are plotted on lognormal probability paper in Figure 7.2. The lines through the points were obtained by using least squares on the equation

$$\ln t = \ln T_{50} + \sigma\Phi^{-1}[F(t)]$$

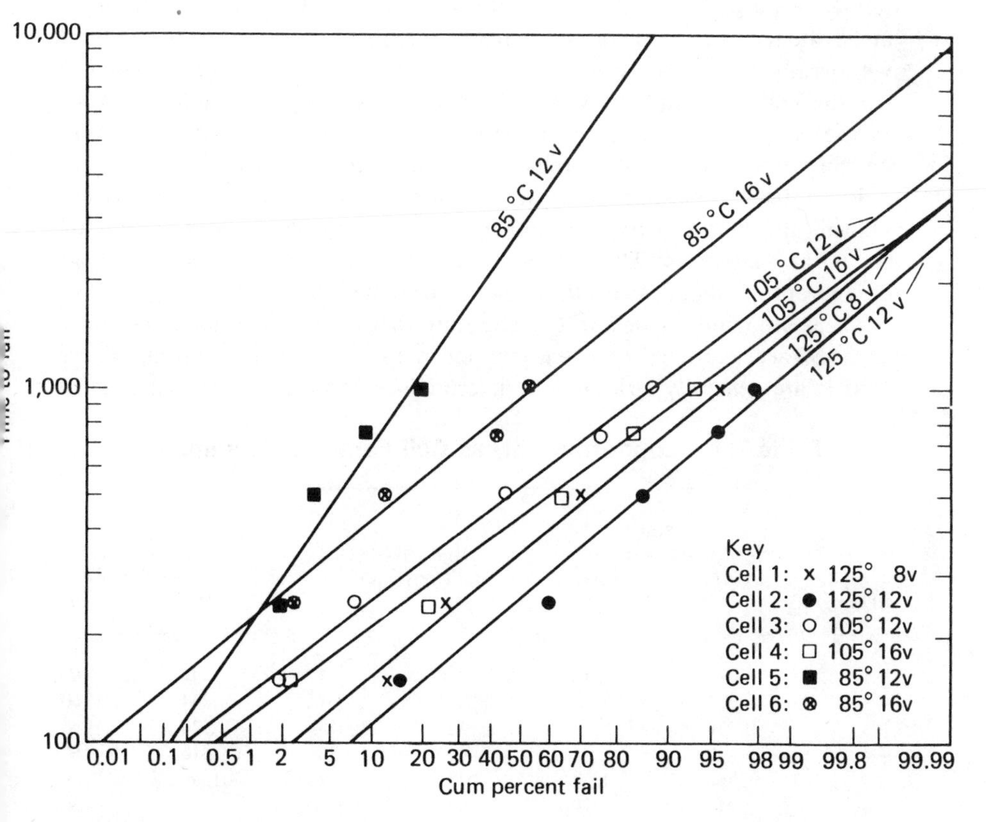

Figure 7.2. Plot of Lognormal Data Cells. Cumulative Percent Fail Data from Table 7.3.

where Φ^{-1} is the inverse of the standard normal distribution and $F(t)$ is the cumulative fraction failed up to time t. Here, the natural logarithm of time is the dependent variable and the normal inverse of the CDF estimate is the independent variable.

It can be seen from the graph that all the lines have very similar slopes; only cell 5, with significantly less data than the other cells, has a somewhat different looking σ. Estimates of the cell T_{50}'s and σ's are given in Table 7.4, using the least-squares graphical method and maximum likelihood estimation (MLE). In addition, constrained MLEs with σ made equal across all cells, are shown. These last estimates are the best possible under the assumption of linear acceleration and lognormality.

ACCELERATION MODELS

From the preceding sections we have seen that knowing the scale parameters (either c or T_{50}) for the life distributions at two stresses allows us to immediately calculate the acceleration factor between the stresses. Alternatively, if we already know the acceleration factor between a laboratory stress test and the field use condition, we can convert the results of our test data analysis to use condition failure rate projections. Indeed, this is often done as an ongoing process monitor for reliability on a lot by lot basis.

But what can be done if an acceleration factor to use conditions is not known, and data can only be obtained in a reasonable amount of time by testing at high stress? The answer is we must use the high stress data to fit an appropriate model that allows us to extrapolate to lower stresses.

There are many models in the literature that have been used successfully to model acceleration for various components and failure mechanisms. These models are generally written in a deterministic form that says that time to

Table 7.4. Lognormal Stress Cell Parameter Estimates.

CELL	GRAPHICAL ESTIMATES (LEAST SQUARES)		MLE ESTIMATES (SEPARATE CELLS)		MLE ESTIMATES (WITH 1 σ)	
	T_{50}	σ	T_{50}	σ	T_{50}	σ
1	340	.62	346	.64	346	.60
2	254	.63	244	.61	244	.60
3	516	.57	523	.55	524	.60
4	422	.54	410	.56	410	.60
5	2760	1.10	2121	.88	1596	.60
6	918	.61	925	.60	927	.60

fail is an exact function of the operating stresses and several material and process dependent constants.

Since all times to fail are random events that can not be predicted exactly in advance, and we have seen that acceleration is equivalent to multiplying a distribution scale parameter, we will interpret an acceleration model as an equation that calculates a distribution scale parameter, or percentile, as a function of the operating stress.

For example, if a failure mechanism depends on two stresses and follows a lognormal model, an equation $T_{50} = G(S_1, S_2)$ that predicts the T_{50} based on the values of these two stresses, is an acceleration model. In the next two sections we shall see several common and useful forms for G.

Before proceeding to these models, however, one point should be stated. Just as, in general, different failure mechanisms follow different life distributions, they may also have different acceleration models. In Chapter 8 we will discuss how, using the competing risk model, we can study each failure mode and mechanism separately, and derive the total component failure rate as a sum of the individual failure rates from each mechanism. This method is virtually the only way to do acceleration modeling successfully.

For example, one failure mechanism might involve a chemical reaction, and be accelerated by temperature. A component with this type of failure mode might also fail due to metal migration, which could be highly dependent on voltage and humidity, in addition to temperature. At the same time there could also be a mechanical wearout failure mode dependent on the frequency of on and off cycles. Each of these modes of failure will follow completely different acceleration models, and must be studied separately.

Therefore, when we discuss acceleration models and how to analyze stress cell failure data, we are presupposing the experiments have been carefully designed to produce data from only one failure mechanism, or any other types of failures have been "censored" out of the data analysis. The way we censor them out is by pretending they were removed from test without having failed. (The time of removal to use is the time they actually failed due to the different failure mechanism.)

THE ARRHENIUS MODEL

When only thermal stresses are significant, an empirical model, known as the Arrhenius model, has been used with great success. This model takes the form

$$T_{50} = Ae^{\Delta H / k\mathrm{T}}$$

where A and ΔH are unknown constants, k is Boltzmann's constant, and T is temperature measured in degrees Kelvin at the location on the component where the failure process is taking place. Boltzmann's constant has the value 8.617×10^{-5} in EV/°K or 1.380×10^{-16} in ergs/°K. Temperature in degrees Kelvin is obtained by adding 273.16 to temperature in degrees Celsius.

Note that we can write the Arrhenius model in terms of T_{50}, or the c parameter (when working with a Weibull), or the $1/\lambda$ parameter (when working with an exponential), or any other percentile of the life distribution we desire. The value of the constant A will change, but as we are about to see, this will have no effect on acceleration factors. For convenience, we will use c's or T_{50}'s in this chapter, but everything applies equally well to other percentiles.

We solve for the acceleration factor between temperature T_1 and temperature T_2 by taking the ratio of the times it takes to reach any specified CDF percentile. In other words, the acceleration factor AF between stress 1 and stress 2 is defined to be the ratio of time it takes to reach $P\%$ failures at stress 1 divided by the time it takes to reach $P\%$ failures at stress 2. The assumption of true acceleration makes this factor the same for all P. Using the Arrhenius model and the fiftieth percentile, we have

$$\mathrm{AF} = \frac{T_{50_1}(\text{at } \mathrm{T}_1)}{T_{50_2}(\text{at } \mathrm{T}_2)} = \frac{Ae^{\Delta H/k\mathrm{T}_1}}{Ae^{\Delta H/k\mathrm{T}_2}}$$

from which

$$\mathrm{AF} = e^{(\Delta H/k)[(1/\mathrm{T}_1) - (1/\mathrm{T}_2)]}.$$

This shows that knowing ΔH alone allows us to calculate the acceleration factor between any two temperatures. Conversely, if we know the acceleration factor, we can calculate ΔH as follows:

$$\Delta H = k\left[\ln\left(\frac{T_{50_1}}{T_{50_2}}\right)\right]\left(\frac{1}{\mathrm{T}_1} - \frac{1}{\mathrm{T}_2}\right)^{-1}.$$

This last equation shows us how to estimate ΔH from two cells of experimental test data consisting of times to fail of units tested at temperature T_1 and times to fail of units tested at temperature T_2. All we have to do is estimate a percentile, such as T_{50}, in each cell, then take the ratio of the corresponding times and use the preceding equation to estimate ΔH. This procedure is valid for any life distribution.

It is useful to have a quick table lookup of acceleration factors as a function of temperatures and ΔH. Table 7.5 shows the acceleration factor from a low T_1 to a high T_2 for two values of ΔH; the acceleration factor if ΔH is .5 is shown in normal print and the acceleration factor for $\Delta H = 1.0$ is in

Table 7.5. Arrhenius Acceleration Factors for $\Delta H = .5$ and ($\Delta H = 1.0$).

LOWER TEMPERATURE T_1 (°C)	HIGHER TEMPERATURE T_2 °C									
	65	75	85	95	105	115	125	135	145	155
25	10.0 (100)	16.4 (268)	26.1 (679)	40.5 (1637)	61.4 (3767)	91.1 (8305)	133 (17,597)	190 (35,938)	266 (70,933)	368 (135,628)
35	5.3 (28)	8.7 (76)	13.9 (192)	21.5 (463)	32.6 (1065)	48.5 (2348)	70.5 (4976)	101 (10,163)	142 (20,059)	196 (38,354)
45	2.9 (8.6)	4.8 (23.2)	7.7 (58.8)	11.9 (142)	18.1 (326)	26.8 (719)	39.0 (1524)	55.8 (3111)	78.4 (6141)	108 (11,743)
55	1.7 (2.8)	2.8 (7.6)	4.4 (19.3)	6.8 (46.6)	10.4 (107)	15.4 (237)	22.4 (501)	32.0 (1021)	44.9 (2020)	62.2 (3865)
65		1.6 (2.7)	2.6 (6.8)	4.0 (16.4)	6.1 (37.7)	9.1 (83.1)	13.3 (176)	19.0 (360)	26.6 (710)	36.8 (1358)
75			1.6 (2.5)	2.5 (6.1)	3.8 (14.1)	5.6 (31.0)	8.1 (65.7)	11.6 (134)	16.3 (265)	22.5 (507)
85				1.6 (2.4)	2.4 (5.5)	3.5 (12.2)	5.1 (25.9)	7.3 (52.9)	10.2 (104)	14.1 (200)
95					1.5 (2.3)	2.3 (5.1)	3.3 (10.8)	4.7 (22)	6.6 (43.3)	9.1 (82.9)
105						1.5 (2.2)	2.2 (4.7)	3.1 (9.5)	4.3 (18.8)	6.0 (36)
115							1.5 (2.1)	2.1 (4.3)	2.9 (8.5)	4.0 (16.3)
125								1.4 (2.0)	2.0 (4.0)	2.8 (7.7)
135									1.4 (2.0)	1.9 (3.8)
145										1.4 (1.9)

parentheses. Tables 7.6 and Figures 7.3 and 7.4 provide a two-step procedure for calculating acceleration factors for a wide range of ΔH. First the T_1 and T_2 values are used to find a *TF* value (defined below) from Table 7.6. Then this *TF* value and the appropriate ΔH line are used in Figure 7.3 or 7.4 to find the acceleration factor. The acceleration factor can also be calculated from

$$AF = e^{\Delta H \cdot TF}.$$

Since the *TF* values in Table 7.6 are just the reciprocal of the last part of the right-hand side of the equation for calculating ΔH ($TF = [(1/T_1) - (1/T_2)] \times 11605$), we can rewrite that equation as

$$\Delta H = \frac{\ln (T50_1/T50_2)}{TF}.$$

This is a convenient form for using Table 7.6 to calculate ΔH when two cells of data have been run.

EXAMPLE 7.4 ARRHENIUS ACCELERATION

What is the acceleration from a use condition temperature of 35°C to a laboratory testing stress of 125°C if ΔH is 1.0? Use the two-step procedure

Table 7.6. TF Values for Calculating Acceleration. Use *TF* Value and Figure 7.3 or 7.4 to Calculate Acceleration.

LOWER TEMPERATURE T_1 (°C)	HIGHER TEMPERATURE T_2 (°C)									
	65	75	85	95	105	115	125	135	145	155
25	4.6	5.6	6.5	7.4	8.2	9.0	9.8	10.5	11.2	11.8
35	3.3	4.3	5.3	6.1	7.0	7.8	8.5	9.2	9.9	10.6
45	2.2	3.1	4.1	5.0	5.8	6.6	7.3	8.0	8.7	9.4
55	1.0	2.0	3.0	3.8	4.7	5.5	6.2	6.9	7.6	8.3
65		.99	1.9	2.8	3.6	4.4	5.2	5.9	6.6	7.2
75			.93	1.8	2.6	3.4	4.2	4.9	5.6	6.2
85				.98	1.7	2.5	3.3	4.0	4.6	5.3
95					.88	1.6	2.4	3.1	3.8	4.4
105						.79	1.5	2.3	2.9	3.6
115							.75	1.5	2.1	2.8
125								.71	1.4	2.0
135									.68	1.3
145										.65

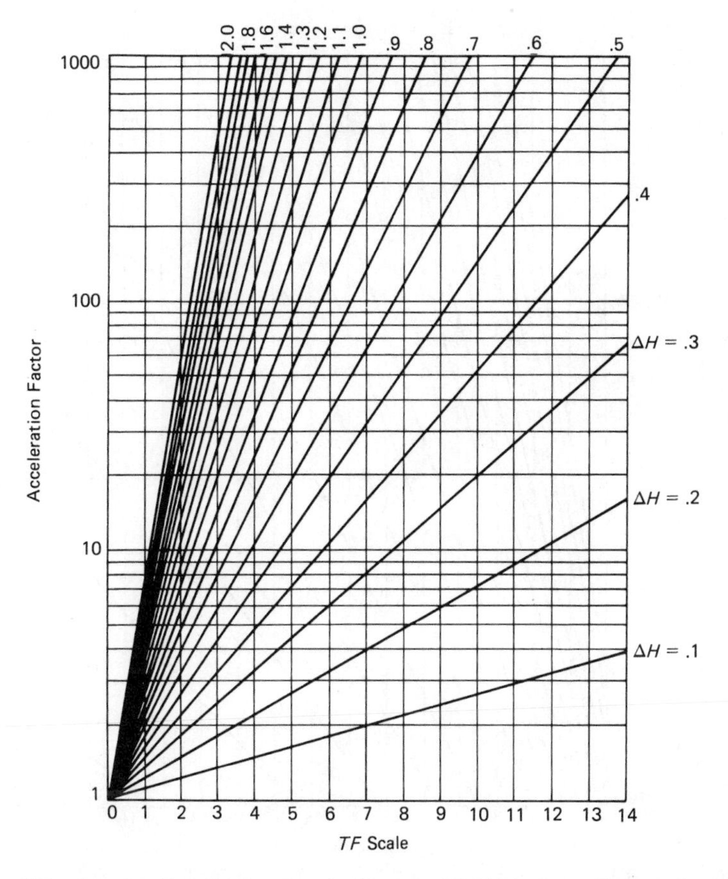

Figure 7.3. Acceleration Factor Graph. Choose *TF* Value from Table 7.6 and Find Acceleration for a Given ΔH.

involving Tables 7.6 and Figure 7.3 or 7.4 to check the result obtained from Table 7.5. What would the acceleration be if ΔH is only .7?

SOLUTION

Table 7.5 gives a direct lookup value for the acceleration factor when ΔH is 1.0 of 4976. Using Table 7.6, a TF value of 8.5 is obtained for the two temperatures. Going up from this value on the TF scale of Figure 7.4 to the line for $\Delta H = 1.0$, and then going over to the vertical acceleration scale, yields the more approximate estimate of 4970. For a ΔH of .7, the acceleration corresponding to the same TF value is 385 (from Figure 7.3).

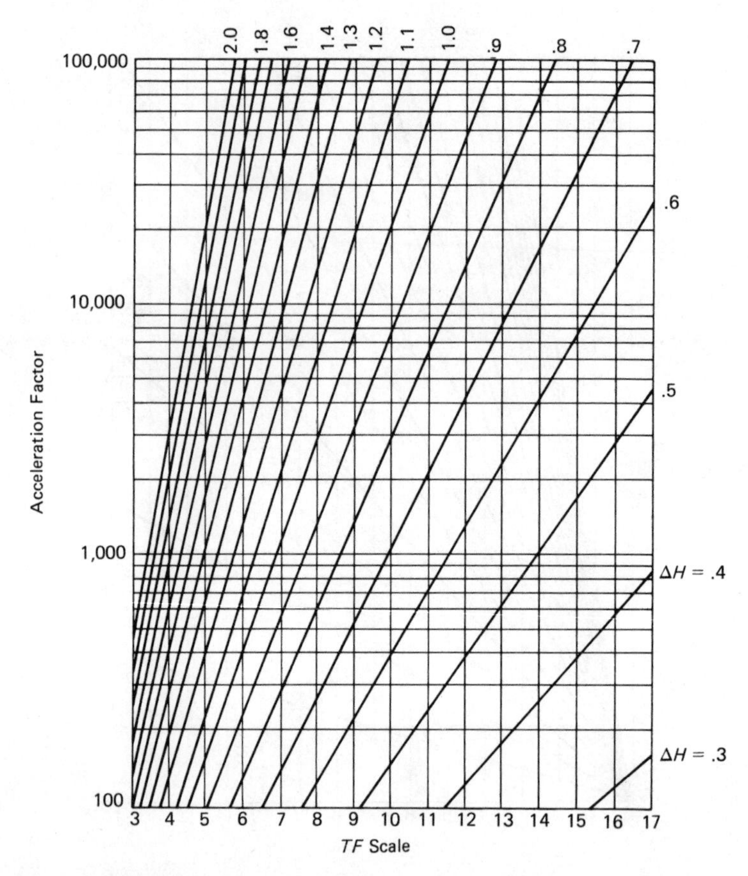

Figure 7.4. Acceleration Factor Graph. Choose *TF* Value from Table 7.6 and Find Acceleration for a Given ΔH.

EXAMPLE 7.5 CALCULATING ΔH WITH TWO TEMPERATURE CELLS

Use the 85°C and 125°C cells of Weibull data from Example 7.2 and Table 7.2 to calculate an estimate of ΔH.

SOLUTION

Using the graphical estimates for c from Example 7.2, and the *TF* value of 3.3 from Table 7.6, we calculate ΔH to be

$$\Delta H = \frac{\ln (40194/242)}{3.3} = 1.55.$$

The same calculation using the generally more accurate MLE values of 9775 and 221 for c estimates yields $\Delta H = 1.15$. Using the constrained equal shape MLE values of 12,270 and 229 gives a ΔH estimate of 1.21. Such differences between methods of estimation are not uncommon and will be seen again in later examples in this chapter.

ESTIMATING ΔH WITH MORE THAN TWO TEMPERATURE CELLS

In the last example, we only made use of the 85°C and the 125°C cells from Example 7.2. If we had used the graphical $\hat{c}$ values from the 85°C and the 105°C cells, we would have calculated a ΔH estimate of (ln 40,194/2208)/1.7 = 1.71. Or, if we had used graphical $\hat{c}$ values from the 105°C and the 125°C cells, the estimate would be (ln 2208/242)/1.5 = 1.47. Which of these three estimates is the right (or best) one to use?

The answer is none of them. We need a procedure that makes use of all the cells simultaneously to derive a ΔH estimate. This can be done either graphically or using regression techniques. The key equation is the Arrhenius model written in logarithmic form:

$$\ln c = \ln A + \Delta H\left(\frac{1}{k\mathrm{T}}\right).$$

This equation is linear in the (independent) variable $1/k\mathrm{T}$ and the (dependent) variable ln c, with slope ΔH and intercept ln A. So a plot of the data points, in the form (ln c, $1/k\mathrm{T}$) from each cell, should line up almost in a straight line. The slope of this line is an estimate of ΔH. Slope (as discussed in Chapter 6) is calculated by taking any two points on the line and calculating the ratio of the increase in height going from the left to the right, over the distance between the points along the x axis (or $1/k\mathrm{T}$ axis). Like semilog paper, units must be adjusted because of the logarithm calculation, so that going one decade on the y axis is equivalent to 2.3 units on the x axis.

Special graph paper can be constructed where the y axis is logarithmic and the x axis is calibrated in terms of reciprocal temperature multiplied by Boltzmann's constant, simplifying the plotting of cell data points. The point for a cell is located by looking up c (or T_{50} for the lognormal) on the y axis, and the cell temperature on the x axis. Various commercial brands of this paper are available, such as Keuffel and Esser.

Once the points are plotted, the best straight line through the points represents the fitted Arrhenius model. It can be used as an instant calculator for finding the c corresponding to any temperature, by going up from the temperature to the line, and then to the left to the c value on the y axis.

The line can be drawn subjectively by eye, or better yet, fitted using least squares on the log equation. By looking at how well the points follow a straight line a subjective evaluation of the appropriateness of the Arrhenius equation is obtained.

EXAMPLE 7.6 CALCULATING ΔH WITH THREE TEMPERATURE CELLS

Use the three Weibull cells of Example 7.2 to obtain a graphical fit of the Arrhenius model equation and an estimate of ΔH using all the cells.

SOLUTION

The points are plotted in Figure 7.5, along with the least-squares (regression) line. For the regression, the independent variable $1/k$T values are 32.40,

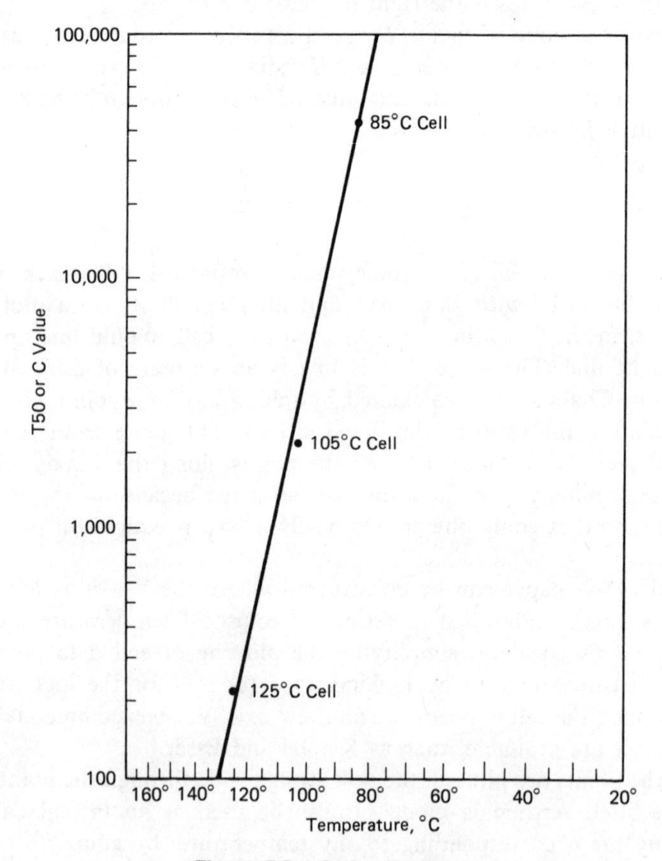

Figure 7.5. Arrhenius Plot.

30.69, 29.15, corresponding to the temperatures of 85, 105, and 125°C. The matching dependent variable values are ln $\hat{c}$ or 10.60, 7.70, and 5.49, corresponding to the graphical $\hat{c}$ estimates of 40194, 2208, and 242. The ΔH estimate turns out to be 1.58. The actual points fit the line very closely.

If a line had been drawn by eye through the plotted data points in Figure 7.5, the slope could be calculated by picking two convenient points at opposite ends of the line and using the formula for ΔH.

Since it is generally more accurate to use maximum likelihood estimates for $\hat{c}$, this would lead to an independent variable with values of 9.18, 7.47, and 7.7. The resulting ΔH estimate is 1.16. Finally, if the still better estimates from constrained maximum likelihood are available (where m is forced equal across the three cells), the estimate of ΔH is 1.22.

A natural question, now that we have analyzed the data from Example 7.2 by several methods and fit an Arrhenius model to it, is how well did we actually do? In a real analysis we may not know the answer to that question for years, if at all. In this case, however, all the data points were obtained by simulation from known Weibull distributions and a known Arrhenius equation. Thus we can display the results of our analysis and compare them to the "real" values. This comparison is shown in Table 7.7. The results show that both the MLE and the constrained shape MLE did roughly as well, and both were considerably better than the completely graphical analysis. This outcome agrees with what we would expect from theory.

The use of graph paper was illustrated for Weibull data and the c parameter in Figure 7.5, but the methodology is really independent of any particular distribution. It can also be applied to the T_{50}, or any other percentile. Since percentiles can be estimated, as described in Chapter 1, without knowing

Table 7.7. Weibull Data Analysis Summary.

	ESTIMATION METHOD			
PARAMETER	GRAPHICAL	MLE	CONSTRAINED MLE	TRUE VALUE
ΔH	1.58	1.16	1.22	1.20
ln A	−40.50	−28.35	−30.06	−29.52
m	.66 (3-cell average)	.77 (3-cell average)	.72	.70
c at 85°C	37,597	10,706	13,708	11,424
c at 105°C	2,542	1,466	1,616	1,461
c at 125°C	224	244	246	230

what the distribution is, it is not necessary to assume a form for the failure mechanism life distribution in order to fit an Arrhenius model. An analysis without assuming any form of distribution is called a nonparametric analysis. What we gain from assuming a suitable distribution model is accuracy (provided the assumption is approximately valid) and the ability to project results to percentiles that can not be observed using the small sample sizes reliability analysts typically work with. Hence parametric models are the general rule.

One last rather sophisticated method for calculating ΔH should be mentioned. It is possible to embed the Arrhenius model directly into the maximum likelihood equations for the Weibull or lognormal or exponential cell data. Then ΔH becomes a parameter that is directly estimated by the MLE method, using data from all the cells. This method requires highly sophisticated computer programs, and will not be mentioned further except to state that such programs are available commercially.

THE EYRING MODEL

The Arrhenius model is an empirical equation that justifies its use by the fact that it "works" in many cases. It lacks, however, a theoretical derivation and the ability to model acceleration when stresses other than temperature are involved. Example 7.3, with both temperature and voltage playing key roles, could not be handled by the Arrhenius model.

The Eyring model offers a general solution to the problem of additional stresses. It also has the added strength of having a theoretical derivation based on chemical reaction rate theory and quantum mechanics. In this derivation, based on work by Eyring, the parameter ΔH has a physical meaning. It represents the amount of energy needed to move an electron to the state where the processes of chemical reaction or diffusion or migration can take place. For further details see Eyring, Glasstones and Laidler (1941).

The Eyring model equation, written for temperature and a second stress, takes the form

$$T_{50} = A\,\mathrm{T}^{\alpha} e^{\Delta H/k\mathrm{T}}\, e^{[B+(C/\mathrm{T})]S_1}.$$

The first exponential is the temperature term, while the second exponential contains the general form for adding any other type of stress. In other words, if a second nonthermal stress was needed in the model, a third exponential multiplier exactly the same as the second, except for replacing B and C by additional constants D and E, would be added to the equation. The resulting Eyring model for temperature and two other stresses would then be

$$T_{50} = \mathrm{A}\mathrm{T}^{\alpha} e^{\Delta H/k\mathrm{T}}\, e^{[B+(C/\mathrm{T})]S_1}\, e^{[D+(E/\mathrm{T})]S_2}.$$

As before, T_{50} can be replaced by c or any other percentile or even the mean time to fail.

It is interesting to look at how the first term, which models the effect of temperature, compares to the Arrhenius model. Except for the T^{α} factor, this term is the same as the Arrhenius. If α is close to zero, or the range over which the model is applied is small, the term T^{α} has little impact and can be absorbed into the A constant without changing the practical value of the expression. Consequently, the Arrhenius model is successful because it is a useful simplification of the theoretically derived Eyring model.

When we try to apply the Eyring model to life test data with several critical stresses, however, we often run into several difficulties. The first of these is the general complexity of the model. The temperature term alone has three parameters to estimate—A, α, and ΔH. Each additional stress term adds two more unknown constants, making the model difficult to work with.

As a minimum, we need at least as many separate experimental stress cells as there are unknown constants in the model. Preferably, we should have several more beyond this minimal number, so that the adequacy of the model fit can be examined. Therefore, a three-stress model should have about 9 or 10 cells. Obviously, designing and conducting an experiment of this size is not simple.

Another difficulty is finding the proper functional form, or units, with which to express the nonthermal stresses. Temperature is in degrees Kelvin. But how should voltage or humidity be inputted, for example? The theoretical model derivation does not specify—so the experimenter must either work it out by trial and error, or derive an applicable model using arguments from physics and statistics.

As an example, consider the temperature and voltage acceleration model given by

$$T_{50} = Ae^{\Delta H/kT} V^{-B}.$$

This expression may not look much like a two-stress Eyring model, but, in fact it is. Make the substitutions $\alpha = 0$, $C = 0$, and $S_1 = \ln V$ (where V is volts), and the two-stress Eyring model reduces to this simple acceleration equation with three unknown parameters to estimate.

EXAMPLE 7.7 TWO-STRESS ACCELERATION MODEL

Use the results of the analysis of the six stress cells of lognormal data from Example 7.3 to find the three unknown parameters in the acceleration model for temperature and voltage given above. Estimate the average failure rate at a typical use condition of 25°C and 4 V.

SOLUTION

There are several ways to proceed, depending on the statistical tools (usually computer programs) available to the analyst. The simplest approach is to look at the three 12-V cells, with temperatures of 125, 105, and 85°C. Since voltage is constant, and the temperature portion of the model is Arrhenius, we can analyze just these three cells with the methods described in Example 7.6. This procedure gives us estimates of A and ΔH. To estimate B we note that there are pairs of cells at each temperature which only differ in the voltage used. Taking the ratio of the T_{50}'s for any pair cancels out the Arrhenius portion of the model and yields a simple equation for B.

For example, using the graphical estimates for the 125°C cells yields $(254/340) = (8/12)^B$. Taking logarithms of both sides and solving for B gives an estimate of .72. The same method with the two 105°C cells yields .70. The 85°C cells give the much higher B estimate of 3.83. An average of these three numbers gives 1.75.

The numbers obtained in this calculation should, however, be disquieting to the analyst. The 125 and 105°C cells yielded consistent estimates in the .7 range. These cells had most of the failures (a total of 188). The 85°C cells, with only 37 failures between them, gave a very different estimate. A practical, ad hoc procedure to correct for this would be to weight the estimates by the numbers of failures and then divide the sum by the total number of failures. This weighted method would result in a B of .92.

A better way to to fit the acceleration model would be to use a multiple regression program on all the cell T_{50}'s simultaneously. First we write the model equation in logarithmic form as

$$\ln T_{50} = \ln A + \Delta H/k\mathrm{T} + B(-\ln V),$$

which can be rewritten in the standard multiple regression format as

$$Y = a + \Delta H X_1 + B X_2$$

after making the obvious substitutions. Then we input the Y, X_1, and X_2 vectors to the regression program to get the model estimates.

The regression procedure can be improved by using weighted regression, with the number of failures in a cell used as the weight for that cell's ln T_{50}.

Once we have estimated the model parameters, straight substitution gives the T_{50} for the use condition of 25° and 4 V. Combined with the best overall sigma estimate, this estimate can be used to calculate the lognormal CDF at 40,000 hr and then the average failure rate over the first 40,000 hr.

As in Example 7.6, all the answers are known for this data because they were simulated from known model parameters and lognormal distributions. Table 7.8 shows the results we would have obtained using regression and weighted regression on graphical or MLE or constrained MLE numbers. The weighted sigma estimates were obtained by weighting the cell sigma estimates shown in Table 7.4 by the cell number of failures. For the constrained MLE method, weights are less important since data from all the cells have

Table 7.8. Lognormal Data Analysis Summary.

PARAMETER	ESTIMATION METHOD			TRUE VALUE
	GRAPHICAL	MLE	MLE CONSTRAINED	
ΔH	.69	.63	.56	
	.57 wtd	.56 wtd	.54 wtd	.55
B	1.57	1.40	1.13	
	1.07 wtd	1.08 wtd	1.06 wtd	1.0
$\ln A$	−10.76	−9.57	−8.07	
	−8.51 wtd	−8.14	−7.76 wtd	−8.09
σ	.68 average	.64 average		
	.62 wtd	.61 wtd	.60	.60
Cell 1 T_{50}	340	346	346	348
Cell 2 T_{50}	254	244	244	232
Cell 3 T_{50}	516	523	524	542
Cell 4 T_{50}	422	410	410	407
Cell 5 T_{50}	2760	2121	1596	1391
Cell 6 T_{50}	918	925	927	1043
25°C, 4 V T_{50}	968,719	489, 323	189,100	
	199,460 wtd	177,187 wtd	151,849 wtd	150,673
25°C, 4 V	.0000035	.00011	.012	
AFR (40,000) in %/K	.012 wtd	.018 wtd	.033 wtd	.034

Weighted (wtd) estimates are those obtained using a weighted regression procedure to fit the acceleration model.

already been pooled to estimate sigma and adjust the T_{50} estimates. A good MLE program will, however, calculate approximate asymptotic variances for the ln T_{50} estimates. The reciprocals of these make appropriate weights, and have been used for this case.

Several very important lessons can be learned from the results in Table 7.8. First note how poorly the straight unweighted graphical analysis did. This result was caused by the high T_{50} estimate in the lowest stress cell (85°C, 12 V), based on only 10 failures (nearly twice the true value). Consequently, we obtained high ΔH and B estimates and a projected use T_{50} more than six times too high. As a result, the use condition average failure rate estimate was optimistic by an incredible four orders of magnitude!

This example underlines the dangers inherent in fitting models and using them to extrapolate way beyond the range of the data. We extrapolate because we have no choice; but we must appreciate the risks involved. If we can be this far off, using simulated data that really follow the assumed models and have no "maverick" points, what can happen with real data?

On the other hand, improving our methods of estimation and model fitting paid handsome dividends. Merely using the cell numbers of failures as weights for a weighted regression program improved the graphical estimation to a bottom line average failure rate estimate less than three times better than the true value. Weighted regression, starting with MLE values, gave a result within two times, while the combination of constrained MLE values and weights based on the asymptotic variance estimates was only off by the negligible amount of 3%.

This improvement, from four orders of magnitude off to only 3%, shows the value of using the best techniques of estimation available. Programs may have to be developed, or purchased, and the analysis becomes more complicated and less intuitive, but the increase in precision is worth the effort. At the very least, using a weighted regression approach with graphical estimates can provide significant improvements. A weighted regression will not let cells with few failures have as much influence as cells with better data.

It should be recalled, however, that estimating distribution and model parameters from random data carries no absolute guarantees that apply on a case by case basis. If we use larger sample sizes, we will generally have more accurate results; the same is true with using "better" statistical techniques. The only assurance we have is that in the long run, over many sets of data, the better technique will prove itself. As an example of how in any given analysis the methodology might not matter, the reader might want to try redoing the Weibull/Arrhenius analysis summarized in Table 7.7 using weighted regression. The ΔH estimate improves only slightly, going from 1.57 to 1.53.

OTHER ACCELERATION MODELS

There are many other models, most of which are simplified forms of the Eyring, which have been successful. A model known as the power rule model has been used for paper impregnated capacitors. It has only voltage dependency, and takes the form AV^{-B} for the mean time to fail (or the T_{50} or the c parameter). This is similar to the model discussed in Example 7.7, without the temperature term.

Another way to model voltage is to have a term such as Ae^{-BV}. This kind of term is easy to work with after taking logarithms.

Humidity plays a key role for many failure mechanisms, such as those related to corrosion or ionic metal migration. The most successful models including humidity have terms such as $A(\text{RH})^{-B}$ or $Ae^{-B(\text{RH})}$, where RH is relative humidity.

A useful model for electromigration failures uses current density as a key stress parameter

$$T_{50} = AJ^{-n} e^{\Delta H/kT},$$

with J representing current density. This mechanism produces open short failures in metal thin film conductors owing to the movement of ions toward the anode at high temperature and current densities. A typical ΔH value is .5 EV, While $n = 2$ is common. The lognormal life distribution adequately models failure times, with σ normally in the .5 to 1.5 range.

Models for mechanical failure owing to cracks and material fatigue or deformation, often have terms relating to cycles of stress or changes in temperature or frequency of use. The (modified) Coffin–Manson model used for solder cracking under the stress of repeated temperature cycling (as an electronic component is powered and unpowered) is an example of such a model. As published by Landzberg and Norris (1969), this model takes the form

$$N_f = Af^{\alpha} (1/\Delta \text{T})^{\beta} G(\text{T}_{\max}),$$

where N_f is the number of cycles to a given percent failures, f is the cycling frequency, ΔT is the temperature range, and G is a factor that depends on the maximum temperature reached in a cycle.

A trial and error approach to model building, using something like a full Eyring and working down, based on fitting to data, should only be tried as a last resort. An experiment based on this approach is likely to be costly and unsuccessful. It is much better to have a model in mind before designing the experiment. This model would either come from a theoretical study of the mechanism or a search of what has been used in the literature. The simplest model that can be found or derived, should be used for as long as it matches experimental data and makes predictions that have not been contradicted by experience.

DEGRADATION MODELS

In the preceding examples, many test failures in several different stress cells were needed to estimate the life distribution and model parameters. In certain cases, however, it is possible to do an analysis even without actual failures. This occurs when there is a measurable product parameter that is degrading over time toward a level that is defined to be a "failure" level.

For example, a component may start test with an acceptable resistance value reading. Over time the resistance reading "drifts." Eventually it reaches a certain unacceptable value, or undergoes an unacceptable percent change, and the part is considered to have failed. At every test readout, the resistance reading can be measured and recorded, even though failure has not yet occurred. If we call each measurement a data point, a test cell of n components yields n data points at every readout time, even though few or none of the components may fail during the test.

We need one key assumption to make use of this data. There has to be some function of the measurable parameter that is changing linearly with time. (In some cases, it is more convenient to have the change be linear with log time or time to an exponent, but we will restrict ourselves to the simple case in this treatment.)

If we call the measurable product parameter Q, and the appropriate function $G(Q)$, then we can write

$$G(Q) = G(Q_0) + R(S)t,$$

where t is time and $R(S)$ is the degradation slope at stress S and Q_0 is the time zero parameter value.

We will survey two ways we can use this equation to fit acceleration models.

Method 1

This method is similar to other previously described graphical techniques in that it is informal, but easy to understand and carry out without sophisticated statistical tools.

On a sheet of graph paper, let the y axis represent $G(Q)$ and the x axis time. Draw a horizontal line across the paper at the $G(Q)$ value that is defined to be a failure. Next plot the data points corresponding to the $G(Q)$

readings for a test unit at each readout. If the right $G(Q)$ has been chosen these points should line up approximately on a straight line. Fit a line through the points, either by eye, or by means of a regression program. Then extend the line until it crosses the failure line and note the corresponding time. This is the derived time of fail for the first unit.

Repeat this procedure for every unit in every stress cell. When this part of the analysis is complete, the derived failure times make up a data set similar in appearance to a life test experiment where every unit was tested until failure. These derived failure times can be used to estimate life distribution parameters. Then we can fit an acceleration model, using the methods we have already discussed.

One problem with this procedure is that readout measurement errors tend to introduce additional variability into the data inflating the measurement of the life distribution shape parameter. For this reason, a σ or m measurement based on actual failures in one high stress cell, is advisable.

Method 2

First we place the additional restriction that $G(Q_0)$ be zero. Typically, this assumption is valid since the most common functions for $G(Q)$ are percent change or absolute change from the initial time zero value. Call the failure level value for $G(Q)$ D (for "distance" to go until failure is reached). The model now is of the form $D = R \times t_f$, or, rate of degradation times time to failure equals distance to failure. Solving for time to failure, $t_f = D/R(S)$. Consequently, time to failure is proportional to the reciprocal of the stress dependent slope. But, as we saw in the preceding sections, time to failure is also proportional to the acceleration model equation. In particular, for an Arrhenius temperature dependent mechanism, we set the slope of degradation equal to the reciprocal of the Arrhenius equation obtaining

$$R(\mathrm{T}) = Ke^{-\Delta H/k\mathrm{T}},$$

where K is a constant term.

In the T_1 temperature cell, we obtain estimates of $R(\mathrm{T}_1)$ by taking $G(Q)/t$ for every unit and every readout time. Label these estimates Y_{11}, Y_{12}, . . ., Y_{1n}. Similarly, in the T_2 temperature cell, we calculate Y_{21}, Y_{22}, . . ., Y_{2n} as estimates of $R(\mathrm{T}_2)$. In general, the jth $G(Q)$ measurement in the ith temperature cell, divided by the readout time, is an estimate of $R(\mathrm{T}_i)$. Note that in this notation, the index n is the number of units in a test cell multiplied by the number of readout times. In the analysis that follows, it makes no difference whether n is a constant, or varies from cell to cell.

The data can be transformed into a simple regression model by taking logarithms of the Y_{ij}. Let $Y_{ij} = \ln Y_{ij}$, $a = \ln K$, $b = -\Delta H$, and $x_i = 1/kT_i$. The model can then be written as

$$Y_{ij} = a + bx_i.$$

Standard regression programs can be used to estimate a and b, with confidence bounds. A use T_{50} can then be projected by dividing D (the "distance" to fail) by $R(T_{use})$. As was the case with method 1, a use shape parameter should be estimated from actual failure data.

More complicated acceleration models like an Eyring for several stresses are handled the same way. The cell slope estimates measuring the rate of degradation are set equal to the reciprocal of the acceleration model equation. This expression for $R(S)$ can usually be transformed into a linear form by taking logarithms and changing variables. A program for multiple regression can then be used to solve the resulting equations.

Using degradation data gives us many data points, even from small numbers of units on test. We can also include stress cells that are close to use conditions, as long as the amount of parameter drift we are measuring stands out from the instrument measurement error. We are protected from misjudging the proper sample sizes and stress levels needed to obtain adequate failures. These are several very desirable properties of degradation or drift modeling.

The disadvantage of using this data to model acceleration, is that it takes us one step further away from reality when we deal with parameter drift instead of actual failures. What do we do about units that do not appear to drift at all? What about those that degrade, and then seem to improve or recover? All these situations present both mathematical and conceptual difficulties that can cast doubt on the analysis validity.

Our recommendation is to use drift or degradation analysis only when the drift mechanism is understood and directly relatable to actual failures. Even then, plan to have at least one cell with high enough stress to produce actual failures. This cell can be used to test the validity of the degradation modeling T_{50} projections.

STEP STRESS DATA ANALYSIS

Another technique used to ensure enough failures when conducting stress tests, is to increase periodically the stress within a cell until almost all of the units have failed. The primary drawback to changing stresses by steps, while an experiment is running, is the difficulty of analyzing the resulting data and constructing models. This section will briefly describe a way of conducting an Arrhenius model analysis, using only one cell of units and periodically increasing that cell's operating temperature.

The method relies on repeated use of the concept that changing stress is equivalent to a linear change in the time scale. If we knew the value of ΔH, we could calculate acceleration factors that relate time intervals at several temperatures, to equivalent time intervals at one fixed reference temperature. The step stress experiment would then be reduced to a single stress experiment, with artificial, calculated time intervals for readouts.

To show how this works, assume the actual readout times are R_1, R_2, . . ., R_k and the test ends at R_k. At each readout time, new failures are recorded and the temperature of the cell is increased. Let the temperature during the interval 0 to R_1 be T_1. The temperature between R_1 and R_2 is T_2, and so on, with the final (highest) temperature T_k occurring during the kth interval between times R_{k-1} and R_k. At readout time R_i, the number of new failures observed is $F_i \geq 0$. N units start test and the survivors at the end of test are $S = N - \sum_{i=1}^{k} F_i$. The real time events of this experiment are diagrammed in Figure 7.6.

Now assume we know ΔH. Then we can normalize all the time intervals to equivalent times at the first temperature T_1. First we calculate all the acceleration factors between later temperatures T_j and the starting temperature T_1. These are given by

$$A_j = e^{(\Delta H/k)[(1/T_1)-(1/T_j)]}.$$

The transformed length of the ith interval becomes $A_i \times (R_i - R_{i-1})$ and the equivalent single stress T_1 readout times are

$$R_1 = R_1$$

$$R_i = R_1 + \sum_{j=2}^{i} A_j (R_j - R_{j-1})$$

$$= R_{i-1} + A_i (R_i - R_{i-1}).$$

Using these transformed readout times, and the numbers of observed failures, lognormal or Weibull or exponential estimation can be carried out.

Of course, the problem with all this is we do not know ΔH in advance. If we did we would not have to run a step stress experiment. So the trick

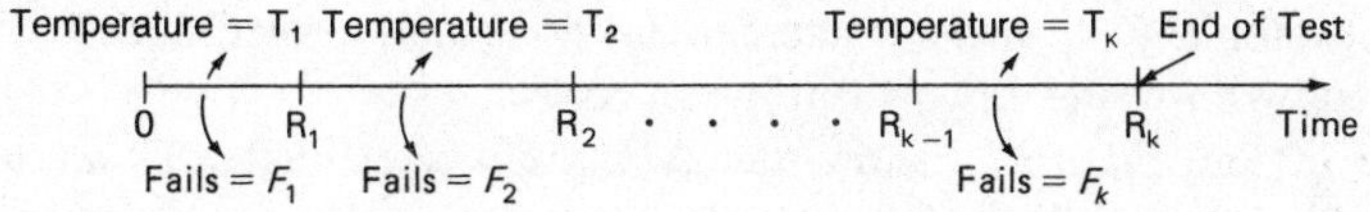

Figure 7.6. Arrhenius Step Stress Data Schematic.

is to keep assuming different values for ΔH, and fitting life distribution parameters to the resulting transformed data. The "best" ΔH is the one where the transformed data most closely fit the assumed life distribution. If maximum likelihood estimation is used, the MLE for ΔH is the assumed value that yields the highest likelihood in the analysis. If graphical techniques are used, the best ΔH gives points that line up closest to a straight line on the appropriate life distribution graph paper.

It is obvious from the above discussion that analyzing data from a multiple stress cell where several different stresses are increasing would be very difficult. A procedure would have to be specially worked out for the assumed acceleration model and the step stress levels.

EXAMPLE 7.8 STEP STRESS MODEL

A sample of 250 components from a population with a temperature dependent lognormal life distribution was tested to determine a T_{50} and σ at any one temperature and an estimate of the Arrhenius ΔH parameter. Since only one test chamber was available, a step stress method was chosen. The initial temperature was 85°C. Readouts were made at 500, 1000, 1500, and 2000 hr. There were no failures at 500 hr; 1 failure at 1000 hr; 7 failures at 1500 hr; and 5 failures at 2000 hr. At 2000 hr, the temperature was increased to 100°C. There were 2 failures at a readout at 2100 hr. Then the temperature was increased to 115°C. The next readout, at 2200 hr, showed 4 failures. At this point a final temperature increase to 125°C was made. The last three readouts, at 2300, 2400, and 2500 hr, showed 36, 38, and 29 failures, respectively. At 2500 hr the test ended, with 128 unfailed units. What are the lognormal parameter and ΔH estimates?

SOLUTION

The iterative MLE method described in this section was used. The results were at 85°C, $T_{50} = 8506$, and $\sigma = .80$ (σ is the same for any temperature). The estimate of ΔH is .86.

Since the data were simulated, we know the "true" values to compare to these estimates. These were $T_{50} = 8000$, $\sigma = .9$, and $\Delta H = .85$. The estimation, especially for ΔH, was fairly close.

Most readers will not have a program capable of this kind of analysis, so this example just shows what can be done and provides check values for anyone who writes his own program to use.

Table 7.9 and Figure 7.7 show how ΔH could be estimated by an iterative graphical procedure. Table 7.9 gives the actual readout times and cumulative percent failures, as well as equivalent readout times (referenced to the 85°C

Table 7.9. Step Stress Example Data.

READOUT TIMES	EQUIVALENT READOUT TIMES $\Delta H = .5$	$\Delta H = .86$	$\Delta H = 1.0$	CUMULATIVE PERCENT FAILURES
500	500	500	500	0
1,000	1,000	1,000	1,000	.4
1,500	1,500	1,500	1,500	3.2
2,000	2,000	2,000	2,000	5.2
2,100	2,192	2,307	2,368	6.0
2,200	2,542	3,168	3,591	7.6
2,300	3,051	4,812	6,184	22.0
2,400	3,560	6,455	8,776	37.2
2,500	4,069	8,099	11,368	48.8

temperature) assuming values of .5 and .86 and 1.0 for ΔH. Figure 7.9 shows a plot on lognormal probability paper of the cumulative percent failures versus the equivalent readout times for each choice of ΔH. The MLE of $\Delta H = .86$ gives the best line, with the high and low ΔH value lines fanning out to the right and the left of the "correct" line. Fitting successive least-

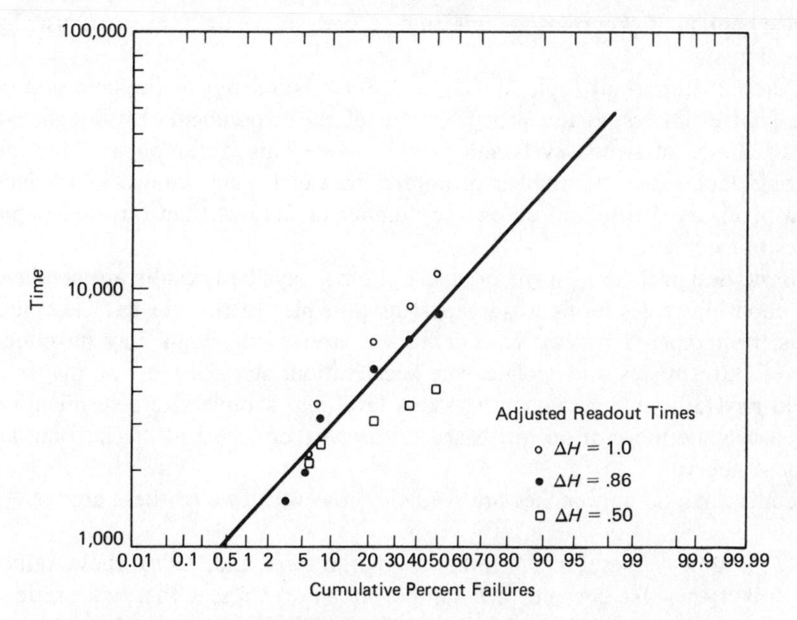

Figure 7.7. Plot of Step Stress Data for $\Delta H = .5$, .86 and 1.0.

squares graphical lines to a range of ΔH values and picking the ΔH that gives the minimum residual line would be the graphical estimation procedure.

One last comment on the design of the experiment described in this example. The most accurate results are obtained by staying at a low level stress as long as needed to get some data, then going quickly to the highest possible stress—as was done in this case.

CONFIDENCE BOUNDS AND EXPERIMENTAL DESIGN

So far in this chapter, only point estimates of acceleration model parameters and use condition failure rates were mentioned. In order to understand the precision of the data and methods, and do proper risk assessment, it is important to have at least approximate confidence intervals for these estimates. In Chapter 3, when we discussed the exponential distribution, it was easy to calculate upper and lower bounds. When we deal with more complicated distributions, and several stress cells of interval or multicensored data, confidence level calculations become much more difficult.

In particular, the quick and easy graphical methods do not allow any confidence interval estimation. Good regression programs will give confidence bounds on acceleration model parameter estimates, and if weighted regression is used, these bounds will be accurate. The asymptotic theory of maximum likelihood estimation offers the best general approach for calculating confidence bounds. If programs for this kind of analysis are used, they will usually give bounds.

Another important topic, having a direct relationship to the precision of the final estimates, is the proper design of the experiment. By design, we mean choice of stress levels and sample sizes. This is complicated for life testing data, since the number of failures in a cell is not known in advance and precision depends more on the number of failures than on the sample sizes put on test.

Basic design theory, in the nonlife test case, says best results are obtained by choosing stress levels as far apart as possible. In the life test case, this must be tempered by two considerations: stresses too high may introduce new failure modes and violate true acceleration; stresses too low may not yield any failures. Consequently, stress level and sample size determination often become more of an art, based on past experience and "feel," than an exact science.

Some analytic approaches are possible, however. Two of these are:

1. Assume values for the unknown parameters and, using these values as if they are correct, pick sample sizes and stresses that will produce an adequate number of failures in each cell. Typically, at least 10 and

preferably over 20 failures in each cell, is adequate. If the data are readout data, failures must be spread out over three or more intervals. Assumed values generally come from typical literature ΔH's and shape parameters. Scale constants such as use T_{50} are calculated by assuming the population just meets its use condition failure rate objective.

2. Make the same unknown parameter value assumptions as in 1. Then simulate cells of failure data for a given experimental design and analyze the data. Repeat this many times to get a feel for the precision of the results. Try again with a different design and see what results are obtained. By iterating on this procedure, a good design will be determined.

Both these methods rely on a good pre-guess of the true model parameters. Kielpinski and Nelson (1975) have carried this approach even further for the lognormal and Meeker and Nelson have studied the Weibull (1975). The have derived optimal life test schemes assuming exact time of failure data and pre-guessed acceleration model and distribution shape parameters.

Example 7.9 shows how a method 1 analysis might be carried out.

EXAMPLE 7.9 PLANNING ACCELERATION EXPERIMENTS

We are testing a new component to verify it meets a use average failure rate objective of .001%/K over 100K hr. The failure mode is temperature dependent and is believed to follow an Arrhenius model. Use temperature is 25°C. We would like to test at 65, 85, and 125°C for 2000 hr. How many units should be put in each cell, and is our choice of temperatures and test time reasonable?

SOLUTION

Assume for the failure mode, a lognormal distribution is appropriate. Our pre-guess for σ is 1, and for ΔH is .7. The average failure rate requirement implies that $F(100\text{K})$ is approximately .001, allowing us to calculate a use T_{50} from

$$T_{50} = 100{,}000\, e^{-\sigma\Phi^{-1}(.0001)} = 100{,}000\, e^{-\Phi^{-1}(.001)}$$

where Φ^{-1} is the inverse of the standard normal distribution, and can be looked up in normal tables. The result is a T_{50} of 2,198,840.

Now we use the ΔH of .7 to calculate acceleration factors between 25 and 65°C or 85 or 125°C. These turn out to be 25, 96, and 937. That leads to T_{50}'s of 87,600 and 22,900 and 2347 for these temperature cells. With such a high T_{50} in the 65°C cell, it is very unlikely we would see any failures at this stress unless we put many thousands of units on test. In the 85°C cell, if we go out to 5000 hr, we would expect 6% of the units to fail. So we modify our design to start at 85°C and place 200 units on test in this cell for 5000 hr.

Since we have decided against the 65°C cell, we could add a cell at 100°C. The acceleration from 25 to 100°C is 239, so the T_{50} is 2,198,842/239 = 9200. This cell will have 27% failures expected at 5000 hr; so a sample size of 100 is adequate.

Finally, we calculate that the 125°C cell will have an expected 44% failures in 2000 hr. One hundred units on test in this cell, for only 2000 hr, will be sufficient.

Note how, if the pre-guessed values are close to the real parameter values, the original experimental design would have yielded insufficient data. An easy calculation ahead of time corrected the situation.

SUMMARY

Many important concepts and techniques were introduced in this chapter. Some were surveyed briefly, while others were developed in detail, by means of examples. These ideas should be studied carefully, for they are typical of situations often encountered in the analysis of life test data.

The theory of acceleration followed from equating "true" acceleration with a linear change in the time scale. From this it followed that lognormal or Weibull shapes remained constant from stress condition to stress condition. Acceleration factors could be calculated by taking the ratios of T_{50}'s, or any other percentile, or even the mean time to fail. Acceleration models describe how the time scale changes as a function of stress. These models can be set equal to sample estimates of some convenient percentile, in order to solve for the unknown model parameters and project to a low stress application.

The following seven steps describe how an acceleration model study might be carried out.

1. Choose as simple an acceleration model for the failure mechanism under investigation as seems appropriate from past experience, or a literature study, or a theoretical derivation. Many models have an Arrhenius temperature term and are simplified Eyring equations.
2. Design an experiment consisting of enough different stress cells to estimate the model parameters. Make sure sample sizes and stresses are such that each cell has an adequate number of failures. More than 10 failures spread out over at least three readout intervals is a rule of thumb for adequate data from a cell. An assessment of proper sample

sizes and stresses can often be made by working backward from the use failure rate objective, and guessing at reasonable model and shape parameters.

3. Choose an appropriate life distribution for the failure mechanism.
4. Analyze the failure data in each cell with as accurate a technique as is available. In any case, do a graphical analysis as well, in order to see how well the data fit the life distribution and follow the equal slope consequence of true acceleration.
5. Fit the acceleration model parameters, again using the best technique available.
6. Make sure that weighted regression is used, especially if the numbers of failures differ widely from cell to cell. The numbers of failures associated with each T_{50} or c estimate make good weights, except when constrained maximum likelihood estimation is used. In that case, use the reciprocal of the asymptotic variance estimates, or if these are not calculated, use no weights.
7. Substitute the estimated model parameters into the model equation, along with use stresses, to project a T_{50} or c. The use shape parameter is the best single value fitting all the stress cells. Obtain the shape estimate as a weighted average of cell shape estimates, using cell failure numbers as weights, or by using constrained maximum likelihood estimation.

Some additional methods that offer the potential of much useful data, with a minimum of testing, are degradation modeling and step stress testing. These procedures require additional assumptions, however, and the resulting data may present analysis difficulties.

8

System Models and Reliability Algorithms

Earlier chapters have described how to estimate reliability distribution parameters and probabilities for components operating at typical use conditions. How can these probabilities be used to predict total system performance? How does the design of the system affect reliability? What are the benefits of redundant design? How can we work backward from an observed system renewal rate to the underlying component CDF?

This chapter will answer questions like these. We will also look at how we can put together component models to form reliability algorithms that provide targets against which actual performance can be measured. In addition, some useful models for defect discovery and early so-called "weak unit" failures will be described.

SERIES SYSTEM MODELS

The most commonly used model for system reliability assumes that the system is made up of n independent components which all must operate in order for the system to function properly. The system fails when the first component fails. This model is called a series or first fail or chain model system. Even though either the independence assumption or the first fail assumption may not be strictly valid for an actual system, this model is often a reasonable and convenient approximation to reality.

In the section on "Some Important Probabilities" in Chapter 2, we derived the formula for the reliability of a series system composed of n identical elements, using the multiplication rule for probabilities. Now we generalize that case to a series system of n, possibly all different, components.

Let the ith component have the reliability function $R_i(t)$. Then the probability the system survives to time t, or the system reliability function $R_s(t)$, is the probability that all the components simultaneously survive to time t. Under the independence assumption, this probability is the product of the individual probabilities of survival (multiplication rule). These probabilities are just the $R_i(t)$. This shows that

$$R_s(t) = \prod_{i=1}^{n} R_i(t) = R_1(t) \times R_2(t) \times \cdots \times R_n(t)$$

or, in terms of the CDF functions

$$F_s(t) = 1 - \prod_{i=1}^{n} [1 - F_i(t)].$$

For system failure rates, the relationship is even simpler:

$$h_s(t) = \sum_{i=1}^{n} h_i(t)$$

$$\text{AFR}_s(T_1, T_2) = \sum_{i=1}^{n} \text{AFR}_i(T_1, T_2).$$

These equations show that for a series system the failure rate can be calculated by summing up the failure rates of all the individual components. There are no restrictions on the types of distributions involved and the result is exact, not an approximation. The only requirements are the independence and first fail assumptions. The proof of this convenient formula is very easy and is outlined in the next example.

EXAMPLE 8.1 SERIES SYSTEMS

Derive the additivity relationship for series system failure rates by using the fact that the failure rate function can be defined as the negative derivative of the natural logarithm of the reliability function (see the section on "The Cumulative Hazard Function" in Chapter 2).

SOLUTION

We have

$$-\ln R_s(t) = -\ln \prod_{i=1}^{n} R_i(t) = \sum_{i=1}^{n} -\ln R_i(t).$$

Therefore, the failure rate is

$$\frac{d[-\ln R_s(t)]}{dt} = \sum_{i=1}^{n} \frac{d[-\ln R_i(t)]}{dt}.$$

from which the additivity of failure rates immediately follows.

THE COMPETING RISK MODEL (INDEPENDENT CASE)

The series model formulas apply in another important case. A single component with several independent failure modes is analogous to a system with several independent components. The failure mechanisms are competing with each other in the sense that the first to reach a failure state causes the component to fail. The series system probability arguments again apply, and the reliability of the component is the product of the reliability functions for all the failure modes. Failure rates are additive, mechanism by mechanism, to get the failure rate of the component.

The more general competing risk model, where the failure processes for each mechanism are not independent, becomes a separate research study depending on how the mechanism random times of failure are correlated. This general model will not be treated in this text.

EXAMPLE 8.2 BOTTOMS-UP CALCULATIONS

A home computer has most of its electronics on one board. This board has 16 memory modules, 12 assorted discrete components, and a microprocessor. The memory modules are specified to have an exponential failure rate with $\lambda = .01\%/K$. The discrete components each have a Weibull CDF with $m = .85$ and $c = 3{,}250{,}000$ hr. The microprocessor is thought to have two significant failure mechanisms. Each mechanism was modeled based on accelerated testing designed to cause that type of failure. The results, adjusted to normal use conditions, yielded a lognormal with $\sigma = 1.4$ and $T_{50} = 300{,}000$ for one mode and an exponential with $\lambda = .08\%/K$ for the other mode. Assuming all components and mechanisms operate independently (at least until the first fail), what is the card failure rate at 5000 hr? What is the chance a card has no failures in 40,000 hr?

SOLUTION

First, we derive the microprocessor failure rate by adding the lognormal and exponential competing failure mode failure rates, evaluated at 5000 hr. This sum is 1600 PPM/K. The reliability at 40,000 hr is the product of the exponential and the lognormal reliability functions. The result is .90. For this part of the example, the competing risk model was used.

Using the series model, we add the failure rates of the 16 memory modules together and obtain 1600 PPM/K. The product of their exponential reliability functions at 40,000 hr is .938.

The 12 discretes each have a Weibull failure rate of 700 PPM/K at 5000 hr, adding 8400 PPM/K to the board total. The product of the 12 reliability functions, evaluated at 40,000 hr, is .7515.

The sum of all the failure rates gives 11,600 PPM/K, or 1.16%/K, for the board total failure rate at 5000 hr. The probability the board lasts 40,000 hr without a failure is the board reliability, or the product of the component reliabilities: $.9 \times .938 \times .7515 = .63$.

This last example showed how, starting with individual failure mode models and using the competing risk and series model as building blocks, a bottoms-up calculation of subassembly or system failure rates is done. Since testing on the system level is usually limited due to time and cost constraints, this bottoms-up approach is of great practical value.

PARALLEL OR REDUNDANT SYSTEM MODELS

A system that operates until the last of its components fails, is called a parallel or redundant system. This system model is the other extreme from the series model where all components must work. Parallel systems offer large advantages in reliability, especially in early life. In applications where good reliability and low front end failure rates have higher priority than component cost, designing in redundancy, at least for key parts of the system, is an increasingly often used option. The computer systems on the space shuttle are an example of this: every system is replicated with several backup copies—even including independent versions of key software.

As before, let the ith system component have CDF $F_i(t)$. The probability the system fails by time t is the probability that all the components have failed by time t. This probability is the product of the CDFs, or,

$$F_s(t) = \prod_{i=1}^{n} F_i(t)$$

$$R_s(t) = 1 - \prod_{i=1}^{n} [1 - R_i(t)] = 1 - \prod_{i=1}^{n} F_i(t).$$

Failure rates are no longer additive (in fact, the system failure rate is smaller than the smallest component failure rate) but must be calculated using basic definitions.

EXAMPLE 8.3 REDUNDANCY IMPROVEMENT

A component has CDF $F(t)$ and failure rate $h(t)$. The impact of this failure rate makes a significant adder to a system currently under design. In order to improve reliability, it is proposed to use two of these components in a parallel (redundant) configuration. Show that the improvement can be expressed as a factor k given by

$$k = \frac{1 + F(t)}{2F(t)},$$

where the old failure rate is k times the new failure rate. How much improvement results when $F(t) = .01$, as compared to later in life when $F(t) = .1$ or $F(t) = .5$?

SOLUTION

The CDF of the two components in parallel is $F^2(t)$ and the PDF, by differentiating, is $2F(t)f(t)$. The failure rate of the pair is (leaving out the time variable t for simplicity)

$$h_s = \frac{2Ff}{1 - F^2} = \frac{2Ff}{(1 + F)(1 - F)} = \frac{2F}{(1 + F)} h.$$

This result shows that $h = k \times h_s$, with k as given above.

When $F = .01$, $k = 50.5$, or about a 50 times improvement. When F is .1, k is only 5.5. For $F = .5$, the failure rate improvement drops to 1.5 times. Thus, redundancy makes a large difference early in life when F is small, and much less of a difference later on. The rule of thumb is one gains by a factor of about $(1/2F)$.

This example can easily be generalized as follows: if a single component with CDF F is replaced by n components in parallel, then the failure rate is improved by the factor

$$k = \frac{1 + F + F^2 + \cdots + F^{n-1}}{nF^{n-1}}.$$

There is about $(1/nF^{n-1})$ times improvement in early life.

Note that the parallel model assumes the redundant components are operating all the time, even when there have been no failures. An alternative setup would be to have the redundant components in a backup or standby mode, only being called on to operate when needed. This model will be discussed next.

STANDBY MODELS AND THE GAMMA DISTRIBUTION

We treat here only the simple case where one or more identical units are on hand, to be used only as necessary to replace failed units. No allowance will be made for the failure rate of a switching device—in practice this element would have to be added into the overall system calculation.

The lifetime until a system failure is the sum of all the lifetimes of the original and standby components, as they each operate sequentially until failure. For n components in the original plus standby group, the system lifetime is

$$T_n = t_1 + t_2 + \cdots + t_n$$

with the t_i independent random variables, each having the single component CDF $F(t)$.

For $n = 2$, the CDF for T_n can be derived using the convolution formula for the distribution of the sum of two independent random variables. For this application, the convolution gives

$$F^2(t) = \int_0^t F(u) f(t-u)\, du.$$

If we now add a third component lifetime and do another convolution, we derive F_3, and so on until for F_n we have

$$F_n(t) = \int_0^t F_{n-1}(u) f(t-u)\, du.$$

For complicated life distribution, such as the Weibull or the lognormal, the convolution integrals would have to be evaluated numerically. In the exponential case, however, the calculations are much simpler.

EXAMPLE 8.4 STANDBY MODEL

A subassembly has a high exponential failure rate of $\lambda = 2\%/\text{K}$. As an insurance backup, a second subassembly is kept in a standby mode. How much does this reduce the failure rate when the subassembly CDF is .01? What about when the CDF is .1 or .5?

SOLUTION

Substituting the exponential CDF and PDF into the convolution formula gives

$$F_s(t) = \int_0^t \lambda e^{-\lambda(t-u)}(1 - e^{-\lambda u})\,du$$
$$= 1 - \lambda t e^{-\lambda t} - e^{-\lambda t}.$$

The standby model PDF is the derivative of this, or

$$f_s(t) = \lambda^2 t e^{-\lambda t}.$$

Using .00002 for λ, the time when $F(t)$ is .01 corresponds to 502.5 hr. $F(t)$ is .1 and .5 at times 5268 and 34,657, respectively. Using these times and λ value, we find the improvement over 2%/K [by calculating h_s using $f_s/(1 - F_s)$] turns out to be about 100 times when the CDF is .01, and 10 times when the CDF is .1. The improvement factor is only about 2.4 times when the CDF is .5.

At the early times, the failure rate improvement given by the standby model with two exponential components was twice that given by the parallel model (see Example 8.3).

We can generalize Example 8.4 to the case of an r level standby system of exponential components. The system lifetime PDF can be derived, by repeated convolutions, to be

$$f_s(t) = \frac{\lambda^r}{(r-1)!}\, t^{r-1} e^{-\lambda t}.$$

This expression is known as the gamma distribution. The parameters are λ and r. The PDF has values only for nonnegative t, and λ and r must be positive numbers. The MTTF is r/λ and the variance is r/λ^2.

In our derivation, r can have only integer values (the number of identical exponential components in the standby system). The general gamma distribution allows r to take on noninteger values, and uses the gamma function (described in Chapter 4) to replace factorials. In this form of the PDF

$$f(t) = \frac{\lambda^r}{\Gamma(r)}\, t^{r-1} e^{-\lambda t}.$$

This is a very flexible distribution form and it is often used empirically as a suitable life distribution model, apart from its derivation as the distribution of a sum of exponential lifetimes.

If we have an r standby exponential model, leading to the above gamma distribution, it can be shown using approximations described in Gnedenko et al. (1969) that the improvement factor is approximately $r!$ times greater in early life than the improvement obtained from the parallel system model. For $r = 2$, this gives a two times improvement, as seen in Example 8.4.

Another special case of the gamma is of interest. When $\lambda = .5$ and r is an integer, by substituting $r = d/2$ the PDF becomes

$$f(t) = \frac{(1/2)^{d/2}}{\Gamma(d/2)} t^{(d-2)/2} e^{-t/2}.$$

This is the chi-square distribution with d degrees of freedom, used to obtain exponential confidence bounds in Chapter 3.

COMPLEX SYSTEMS

Models for systems that continue to operate as long as certain combinations of components are operating can be developed with great generality (see Barlow and Proschan, 1975). Here we discuss two types of complex systems: those that operate as long as at least r components (any r) out of n identical components are working, and those that can be diagrammed as combinations of series and parallel (not necessarily identical) components.

The formula for the reliability function when at least r out of n components must work is obtained by summing the probabilities exactly r and exactly $r + 1$ and exactly $r + 2$, and so on, all the way up to exactly n work. These exact cases are all disjoint events and their sum is the probability at least r out of n are working. Each of these probabilities can be evaluated using the binomial formula (see Chapter 9). The result is

$$R_s(t) = \sum_{i=r}^{n} \binom{n}{i} R^i(t)(1 - R(t))^{n-i}.$$

The CDF and failure rate for this model are derived from the reliability function using basic definitions.

Many systems can be broken down into combinations of components or subassemblies that are in parallel configurations, and combinations that are in series. These systems can be diagrammed like an electric circuit, with blocks logically "in parallel" and blocks logically "in series." The system "working" means there is a path for "electricity" to flow from one end of the diagram to the other. The system may or may not actually have electronic parts—or it may have a combination of electronic components and mechanical components. The electric circuit diagram is used only as a convenient device that helps us reduce the system, by successive steps, to simpler systems with equivalent failure rates.

The three steps in this method are:

1. Diagram the system as if it were an electric circuit with parallel and series components and groups of components. Picture each component as a circle and write R_i within the circles of all the components which have that reliability function.
2. Successively reduce combinations of components by replacing, for example, a group of components that are in series by one equivalent component. This will produce a large circle that has the product of the R's from each circle it replaced as its reliability. For components in parallel, the equivalent component has an R calculated using the parallel model formula (1 minus the product of the individual circle CDFs).
3. Continue in this fashion until the entire system is reduced to one equivalent single component whose reliability function is the same as that of the entire original system.

This procedure sounds complicated and arbitrary. Actually, it turns out to be fairly automatic, after a little practice. A few examples will illustrate how it works.

EXAMPLE 8.5 COMPLEX SYSTEMS

A system has five different parts. Three of them must work for the system to function. If at least one of the remaining two components is working, along with the first three, the system will function. What is the system reliability function?

SOLUTION

The analogous circuit is drawn below.

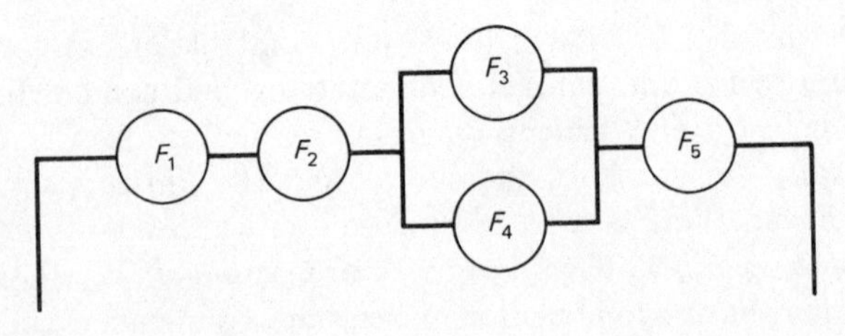

The simplest way to proceed is to replace R_3 and R_4 with one component.

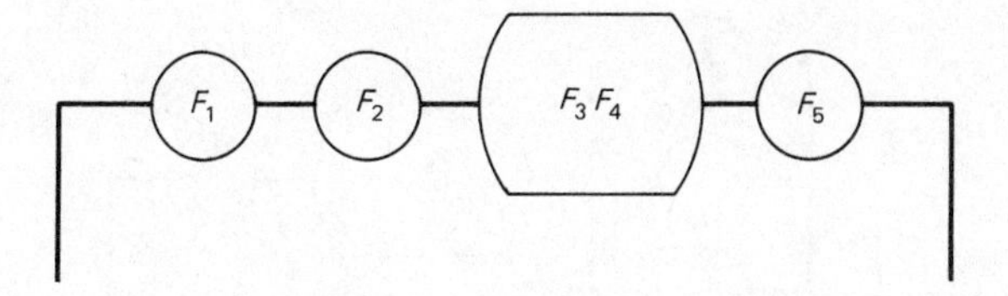

Now all the components are in series and a straightforward application of the series model provides the answer.

$$F_s = 1 - (1 - F_1)(1 - F_2)(1 - F_3 F_4)(1 - F_5)$$

EXAMPLE 8.6 COMPLEX SYSTEMS

A system of six components can be broken up logically into three subassemblies. The first has three components, two of which are the same, and as long as any one of the three works, the subassembly will work. The second subassembly has two different identical components, either of which must work for this part of the system to function. The last (logical) subassembly consists of one critical part. The system works only as long as each subassembly functions. What is the system reliability?

SOLUTION

The diagram for the system, followed by the successive reduction steps solving for the system reliability, are below.

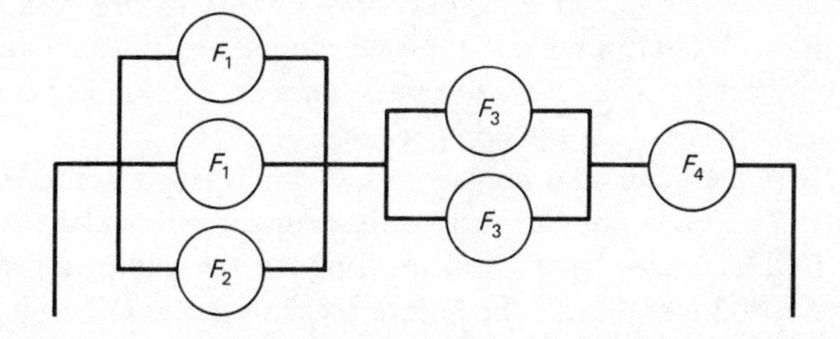

The hardest part of this procedure turns out to be making the initial diagram of the system. Not all systems can be broken down this way, even as an approximation. A simple generalization, left to the reader, would be to allow

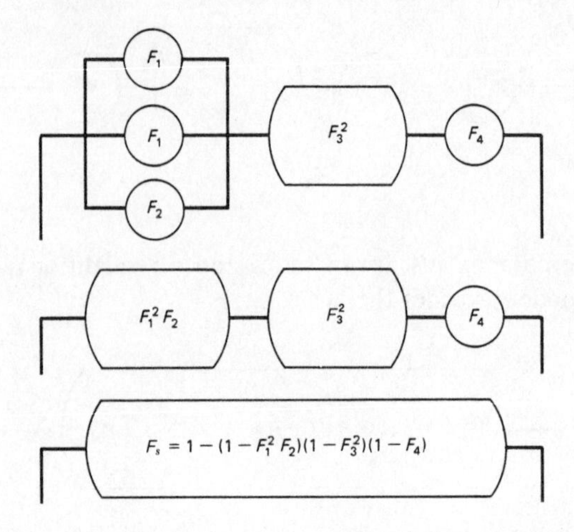

blocks of "r out of n" configured components. The single equivalent component replacing such a block would have a reliability function given by the binomial formula at the start of this section.

DEFECT MODELS AND DISCOVERY DISTRIBUTIONS

This section describes a very useful model that will complete the kit of tools we need to construct very general reliability algorithms that can match the failure rate characteristics of most components and systems encountered in practice.

Suppose a small proportion of a population of components has the property of being highly susceptible to a certain failure mechanism. Consider them to be manufacturing defects in a reliability sense—they work originally and can not be detected as damaged by the standard tests and inspections; yet, after a short period of use, they fail. The rest of the components in the population are either not at all susceptible to this failure mechanism, or else they fail much later in time. A typical example might be a failure mechanism that only occurs when there are traces of a certain kind of contaminant left on critical spots within the component.

Let $F_d(t)$ be the CDF that applies to the small proportion of susceptible components. The CDF for the rest of the components will be denoted by $F_N(t)$. This CDF is usually derived by applying the competing risk model to the life distributions for all the failure mechanisms that apply to normal (nondefective) components.

The total CDF, $F_T(t)$, for the entire population, is constructed by taking a weighted average of the CDFs for the defective and normal subpopulations. The weights are the proportions of each type of subpopulation. This situation is called a mixture. For example, if α proportion are reliability defects, and $(1 - \alpha)$ are normal, then

$$F_T(t) = \alpha F_d(t) + (1 - \alpha)F_N(t).$$

Typically, $F_d(t)$ will be a Weibull or a lognormal with a high probability of failing early, while $F_N(t)$ has a low early failure rate that either stays constant or increases very late in life.

The failure rate of the mixture CDF is

$$h_T(t) = \frac{\alpha f_d(t) + (1 - \alpha)f_N(t)}{1 - [\alpha F_d(t) + (1 - \alpha)F_N(t)]}$$

and this has a "bathtub" shape for the distributions mentioned above.

This model, where a (hopefully small!) proportion of the population follows an early life failure model, appears to have wide applicability. Some ways of spotting it from test or use data will be described later in the chapter (in the section on data analysis). Some authors describe these early failures as "infant mortality," a term that goes back to the actuarial origins of the hazard rate.

The defect mixture model can be extended to cover the almost universal problem of the field discovery of real manufacturing defects that escape in-house detection. No matter how carefully we test components or systems before shipment, some small fraction of defects escapes, to be discovered first by the customer. If this fraction of escapes is α, the mixture model and failure rate equation apply with only a change in the interpretation of $F_d(t)$. This distribution is no longer a life distribution for the failure times of reliability defects. Instead, it is a distribution that empirically models the rate at which customers discover the time zero defects that escape the manufacturing plant. We denote this CDF by $F_e(t)$, and interpret it as a discovery distribution for manufacturing escapes.

A good model for this discovery rate is a Weibull distribution with shape parameter between .5 and .8, and a characteristic life parameter chosen so that 95 or 99% of the defects are discovered by 3000 to 5000 hr. After fine tuning to match the front end curve of actual data, this model can be a standard part of most reliability algorithms.

GENERAL RELIABILITY ALGORITHMS

By starting with a general and flexible form for component failure rates, and using the previously discussed system models to build up to higher assemblies, we can reduce reliability evaluation to an algorithm that handles most applications. The field use CDF would include a term for discovery of escapes, and terms for as many reliability defect subpopulations as are needed. To this would be added the typical population competing failure modes. These competing distributions could have any of the shapes discussed in the chapters on the Weibull and the lognormal distributions (i.e., an early life or a constant or a wearout shape). If we call the CDF for the normal population competing failure modes F_N, the complete algorithm is

$$F_T = \alpha F_e + \beta F_d + (1 - \alpha - \beta) F_N,$$

where F_e is the discovery CDF for the α proportion of defective escapes, F_d is the early life distribution for the β proportion of reliability process defects (if there are more than one type, add as many more such terms as are needed), and F_N is derived from the n normal product competing failure modes as $F_N = 1 - R_1 R_2 \cdots R_n$. This model is most effective if acceleration models for the parameters of F_d and F_N are known. Then the algorithm can be written to give the use CDF as a function of use conditions, making the expression very useful for a product that has varied applications.

In many cases, two CDFs are sufficient to model the normal population competing failure modes. One CDF is an exponential and contributes a constant failure rate. The reliability function for this CDF will be denoted by R_I. The second CDF contributes an increasing or wearout type failure rate. This could be a suitable lognormal or Weibull distribution. The reliability function for this CDF will be denoted by R_w. Then F_N is (by the series model) $F_N = 1 - R_I R_w$. The expression for F_T becomes

$$F_T = \alpha F_e + \beta F_d + (1 - \alpha - \beta)(1 - R_I R_w).$$

EXAMPLE 8.7 GENERAL RELIABILITY ALGORITHM

A dense integrated circuit module consists of an encapsulated microchip connected to a ceramic substrate. After months of life testing and modeling, distributions for the significant failure mechanisms at use conditions are known. There is a constant failure rate of 100 PPM/K and a lognormal wearout distribution with $\sigma = .8$ and $T_{50} = 975{,}000$. In addition, it is estimated that about .2% of the modules have entrapped contaminants that lead to corrosion failure according to a lognormal distribution with $\sigma = .8$ and $T_{50} = 2700$.

If the test coverage and efficiency allows .1% defective modules to be shipped, and the assumed discovery model is a Weibull with $m = .5$ and $c = 400$, give the failure rate curve for the module.

SOLUTION

All the parameters and distributions for the general reliability algorithm have been specified. The value of α is .001, and $\beta = .002$. The use CDF and PDF are calculated from

$$F_T = .001 F_e + .002 F_d + .997(1 - R_I R_w)$$
$$f_T = .001 f_e + .002 F_d + .997(f_I R_w + f_w R_I),$$

where F_e and f_e are the discovery Weibull CDF and PDF, F_d and f_d are the lognormal CDF and PDF for the reliability defects, R_I and f_I are the reliability function and PDF for the exponential (constant failure rate) failure mode, and R_w and f_w are the reliability function and PDF for the wearout lognormal.

A graph of the failure rate $h_T(t) = f_T(t)/[1 - F_T(t)]$ is shown in Figure 8.1. Note the bathtub shape with a steep front end due to the discovery of escapes and reliability defect failures. The failure rate settles down to a little over the constant adder of 100 PPM/K after about 15,000 hr. Wearout starts being noticeable by 50,000 or 60,000 hr, causing the failure rate to rise to nearly 180 PPM/K at 100,000 hr.

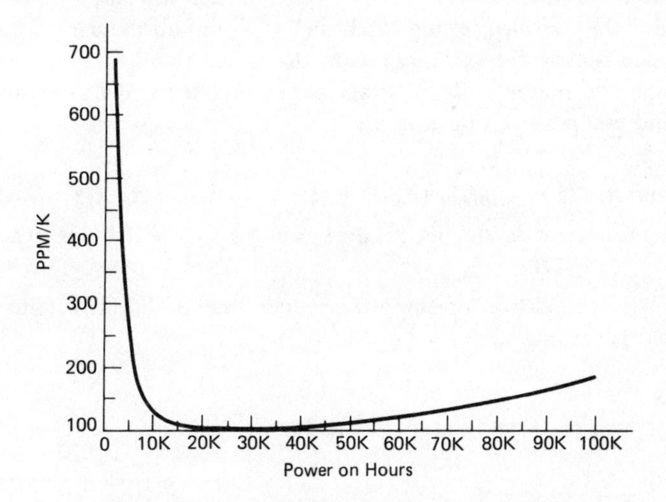

Figure 8.1. Example 8.7 Failure Rate Curve.

BURN-IN MODELS

If the front end of a failure rate curve, such as the one shown in Example 8.7, is too high to be acceptable, and it is not possible to make significant improvements in the manufacturing process, there are two options available to improve the situation. By increasing test coverage and efficiency, the escape discovery portion of the front end can be reduced. Also, by stressing, or "burning in" the product prior to shipment, the reliability defect portion of the front end failure rate can be virtually eliminated. This section discusses how to use our general reliability algorithm to do a mathematical analysis of the possible benefits of a burn-in.

Suppose the burn-in is for T hr at much higher than normal use stress. We need acceleration factors, to use conditions, for every failure mode (apart from defect discovery) in the general algorithm. These factors are obtained using methods such as described in Chapter 7. Assume the factor for the reliability defect mode is A_d, and the factors for the constant and wearout modes are A_l and A_w, respectively. If there are multiple mechanisms (and CDFs) for any of these types, more factors are necessary.

Now we have to decide how efficient the burn-in testing is at catching all the failures produced by the burn-in stress. Are all these failures detected and removed prior to shipment? Or do some escape to be discovered by the end user along with the other manufacturing escapes? Are any previously undetected manufacturing escapes caught by the burn-in testing?

For simplicity, we assume none of the α manufacturing escapes are found and removed at the burn-in. Let the burn-in test efficiency be $(1 - B_e)$. This means that burn-in testing catches $(1 - B_e)$ of all the early life failures that happen during the burn-in period. B_e is the fraction of these failures that escape and merge with the other α manufacturing defect escapes. The fallout and escapes from burn-in are

$$\text{fallout} = (1 - B_e)\{\beta F_d(A_d T) + (1 - \alpha - \beta)[1 - R_l(A_l T)R_w(A_w T)]\}$$

$$\text{new escapes} = \gamma = B_e\{\beta F_d(A_d T) + (1 - \alpha - \beta)[1 - R_l(A_l T)R_w(A_w T)]\}$$

and the new proportion of shipped product that is defective (and will be discovered according to $F_e(t)$) is

$$\text{total escapes to field} = \frac{\alpha + \gamma}{1 - \text{fallout}}.$$

The $(1 - \text{fallout})$ term appears in the denominator because the population actually shipped has been reduced, and the defects are now a higher proportion. Similar correction terms appear in the expression for the field CDF after burn-in. Let

$$\alpha' = \frac{\alpha + \gamma}{1 - \text{fallout}}$$

$$\beta' = \frac{\beta[1 - F_d(A_d T)]}{1 - \text{fallout}}.$$

Then the CDF is

$$F_{BI}(t) = \alpha' F_e(t) + \frac{\beta'[F_d(t + A_d T) - F_d(A_d T)]}{1 - F_d(A_d T)} + (1 - \alpha' - \beta')\frac{1 - R_I(t + A_I T)R_w(t + A_w T)}{R_I(A_I T)R_w(A_w T)}$$

where the correction terms that appear in the denominators can also be viewed as making the probabilities of failure or survival conditional upon surviving the burn-in.

Failure rates are obtained by taking the derivative of the after burn-in CDF $F_{BI}(t)$, and then calculating $h(t)$.

$$f_{BI}(t) = \alpha' f_e(t) + \frac{\beta' f_d(t + A_d T)}{R_d(A_d T)} + \frac{(1 - \alpha' - \beta')}{R_I(A_I T)R_w(A_w T)}[f_w(A_w T + t)R_I(A_I T + t) + f_I(A_I T + t)R_w(A_w T + t)]$$

$$h_{BI}(t) = \frac{f_{BI}(t)}{1 - F_{BI}(t)}$$

EXAMPLE 8.8 BURN-IN MODEL

The reliability engineers responsible for the component described in Example 8.7 decide to try a 9-hr burn-in to improve the front end of the failure rate curve shown in Figure 8.1. The three failure modes described by the CDFs F_d, F_I, and F_w are accelerated by temperature according to an Arrhenius model, with ΔH's of 1.15, .5, and .95, respectively. The field use temperature is 65°C and the proposed burn-in temperature is 145°C. Assuming perfect efficiency at catching failures generated by the burn-in, what will the expected burn-in fallout be? How will the new failure rate curve for burned-in product compare to the old curve? What are the old and new AFR (10,000) values?

SOLUTION

First we use the methods described in Chapter 7 to calculate the Arrhenius acceleration factors. Table 7.6 gives a TF value of 6.6. From Figures 7.3 and 7.4 we read off the approximate acceleration values of $A_d = 2000$, $A_l = 27$, and $A_w = 500$. The fallout at burn-in is

$$\begin{aligned} \text{fallout} &= .002\,F_d\,(9 \times 2000) + .997[1 - R_l(9 \times 27)R_w(9 \times 500)] \\ &= .002 \end{aligned}$$

by substituting the proper CDFs from Example 8.7.

The after burn-in CDF and PDF are

$$\begin{aligned} F_{BI}(t) &= .996966 + .001002\,F_e(t) + .002032\,F_d\,(t + 18{,}000) \\ &\quad - .999R_l(t + 243)R_w(t + 4500) \\ f_{BI}(t) &= .001002 f_e(t) + .002032 f_d\,(t + 18{,}000) \\ &\quad - .999[f_w(t + 4500)R_l(t + 243) + f_l(t + 243)R_w(t - 4500)] \end{aligned}$$

Figure 8.2 shows how $h_{BI}(t)$ compares to the old $h(t)$. The burn-in has made a significant improvement in the front end failure rate, with little effect on the values after about 18,000 hr. The new AFR(10,000) is 120 PPM/K or FIT, as compared to the non-burn-in value of 562 PPM/K. This value was calculated from $\text{AFR}(T) = 10^9 \times [-\ln R(T)]/T$.

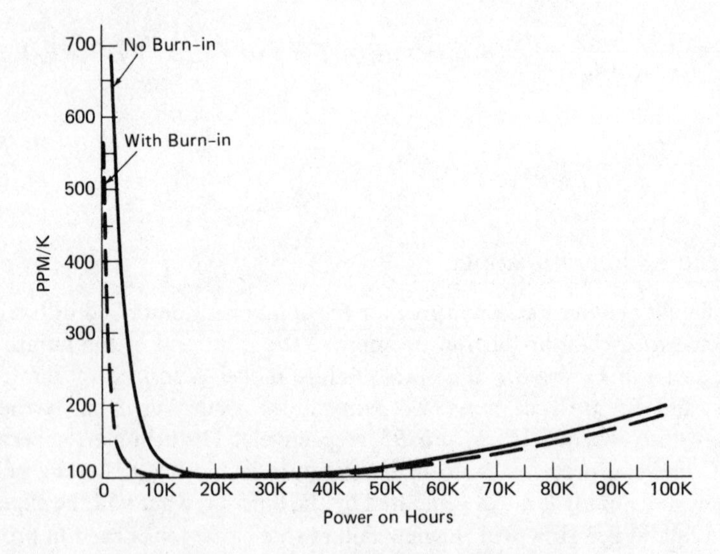

Figure 8.2. Burn-in Improvement Example.

In the last example, the theoretical analysis showed a short burn-in would be beneficial. However, burn-in may not always be so effective since burn-in affects every failure mode differently. Constant failure rate modes are not improved at all, while wearout modes are made worse. Only the early life failure modes, or those due to a reliability defective subpopulation, are helped; and for those to be helped significantly by a short burn-in, there must be a high acceleration factor.

Both the most economical and the most effective way to improve early life performance, is to improve the manufacturing process. With a dedicated return and analysis program of field failures, the major contributors to early life failure can be traced back to mistakes or improvable procedures at specific stages of manufacture. Fixing these improves both yield and reliability.

The next most efficient way to improve early life performance is by more effective tests and screens. Finally, comes the burn-in, or last minute stress screening option. This fix can be thought of as a process bandaid—sometimes necessary no matter how much we would like to avoid it.

The burn-in model of this section can be a useful tool, helping us to choose the best stress level and time for burn-in, or, in some cases, warning us that burn-in is not likely to produce the results desired. Careful data collection, and analysis of both burn-in fallout and later life performance, should be part of any burn-in implementation plan. This is especially true when using stresses of the "shake, rattle, and roll" type, for which no known acceleration model may exist.

Some useful references on burn-in are Peck (1980) and Jensen and Petersen (1982). They call the defect subpopulation failures "freaks" and the escapes infant mortality. They also point out that it is common to introduce infant mortality during the manufacturing operations that occur after a component level burn-in. These added defects may make the burn-in appear less effective.

DATA ANALYSIS

The competing risk model and the defect mixture model have been treated theoretically in the preceeding sections. Here we concern ourselves with how to analyze life test data when one of these models applies.

Assume we have run a life test of a component and are in the middle of plotting the failure data and running analysis programs when we are told by the failure analysis engineers that they can divide the failures into three very distinct modes. Moreover, based on the nature of the failure mechanisms, you might want to use different life distributions to model these modes. How can you statistically separate apart the different failure distributions and estimate their parameters?

This subject was mentioned briefly in the section titled "Failure Mode

Separation" in Chapter 2. Assuming independence of failure modes (i.e., they are not "looking over their shoulder at each other" and changing their kinetics depending on how the other is progressing) and a first fail model, we can treat the data mode by mode as multicensored data. When we are analyzing mode 1, all mode 2 and 3 failures are "censored units" taken off test at the failure time. Multicensored data can be plotted using either hazard plotting procedures or the Kaplan–Meier product probability method (see Chapter 6). Good MLE programs will be able to analytically handle multicensored data and estimate distribution parameters. The key point is to go through the analysis one mode at a time, treating all other modes as units taken off test.

The ability to separate data points by failure mode is critical to the analysis. This separation should be supported with physical analysis; trying statistically to separate failure modes is difficult and dangerous, especially if the distributions have considerable overlap.

Occasionally, a component with several known failure mechanisms has zero failures on life test. Even if we are willing to assume each mode has an exponential life distribution, is there anything we can do with zero failures? Say there are three failure modes with unknown failure rates λ_1, λ_2, and λ_3. Then the component has the constant failure rate of $\lambda = \lambda_1 + \lambda_2 + \lambda_3$, and we can use the zero failures formulas of Chapter 3 to put an upper bound on λ. Unfortunately, this same upper bound applies individually to each of the three mechanism failure rates. So we have the paradoxical result that the upper bound on the total is ⅓ the sum of the upper bounds of the parts. If we have known acceleration factors A_1, A_2, and A_3, for the three modes, the best we can do is use the minimum of these to reduce the component test upper bound to a use condition upper bound.

The final aspect of data analysis, related to the models in this chapter, which we will discuss, is how to detect and analyze data with failures from a reliability defective subpopulation. For example, if we test 100 units for 1000 hr and have 30 failures by 500 hr, and no more by the end of test, are we dealing with two populations or just censored data? If we continue the test will we see only a few more failures, because we have "used up" the bad ones? Or will the other 70 fail according to the same life distribution?

The easiest way to spot that data contain a defect subpopulation is by graphical analysis. Assume, for example, that the failure mode is one typically modeled by a lognormal distribution. If we plot the failures on lognormal graph paper, and instead of following a straight line they seem to curve away from the cumulative percent axis, it is a signal that a defect subpopulation might be present. If we run the test long enough, we would expect the plot to bend over asymptotic to a cumulative percent line that represents the proportion of defectives in the sample.

So the clues are a plot with points that bend and a slowing down or complete stop of failures, even though many units are still on test. In addition, a physical reason to expect reliability defects is highly desirable.

Returning to our 30 failures out of 100 on test illustration, look what happens if we replot the data, this time assuming only 30 units were on test. If these were truly the complete reliability defective population of units on test, and the lognormal model is appropriate for their failure mode, then the new plot should no longer bend. The T_{50} and sigma estimates obtained from the best fitting line (or a MLE program) would be the proper parameter estimates for the $F_d(t)$ of the defective subpopulation. The estimate of β would be .3.

We would run into trouble, however, if there were a number of defective units left on test that have not had enough time to fail. The original plot, with unadjusted sample size, would have less curvature and it might not be clear where the asymptote is. If we worked by trial and error, adjusting the number on test from the full sample size down to the actual number of failures, we might be able to pick one plot where the points line up best. This fitting can be done by computer, using an iterative least-squares fit and choosing the starting sample size that yields the smallest least-squares error.

It is possible to work out more sophisticated techniques (such as MLE) to estimate the proportion of defectives and the distribution parameters. However, no such programs are readily available. Fortunately, practical experience with the simple graphical procedure described above has shown it to work effectively.

EXAMPLE 8.9 DEFECT MODEL

A certain type of semiconductor module has a metal migration failure mechanism that is greatly enhanced by the presence of moisture. For that reason, modules are hermetically sealed. It is known that a small fraction will have moisture trapped within the seal. These units will fail early and it is desired to fit a suitable life distribution model to these reliability defects.

Test parts are made in such a way as to greatly increase the chance of enclosing moisture in a manner typical of the normal manufacturing process defects. One hundred of these parts were put on life test for 2000 hr. There were 15 failures. The failure times were: 597, 623, 776, 871, 914, 917, 1021, 1117, 1170, 1182, 1396, 1430, 1565, 1633, 1664.

Estimate the fraction defective in the sample and lognormal life distribution parameters for this subpopulation.

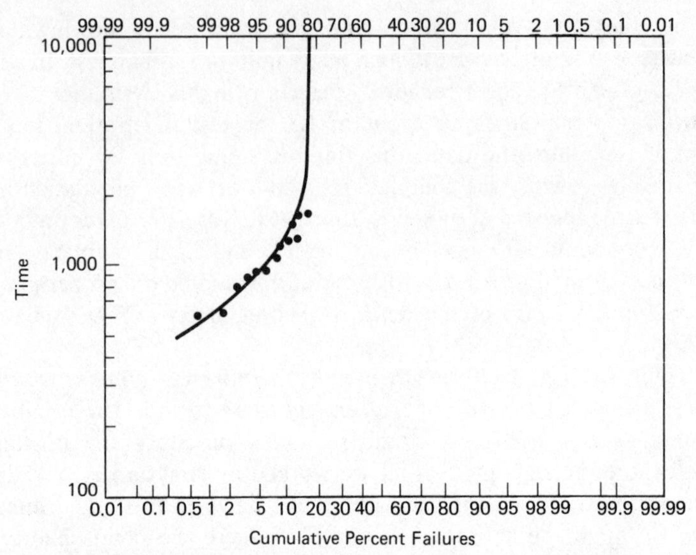

Figure 8.3. Data from Example 8.9 Plotted (Unadjusted).

SOLUTION

Figure 8.3 shows a lognormal plot of the fail times, using $100(i - .3)/100.4$ as the plotting position for the ith fail. The points do appear to have curvature, although exactly where they will bend over to is not readily apparent. By iterative least-squares trials, changing n from 100 down to 15, a "best" fit occurs at $n = 18$. This plot is shown in Figure 8.4. Here the points line up well. The graphical T_{50} and sigma estimates are 1208 and .43. (These results are quite good since the data were simulated from 20 defects with a T_{50} of 1200 and a sigma of .45.)

A final word of caution: although the defect model is very useful and applies in many applications, a slight appearance of curvature on a probability plot does not automatically indicate its presence. Random data can often give the appearance of curvature, especially in the tails, even when all the samples belong to the same population. A long period without further failures is a more reliable clue, and a physical explanation should always be sought. The iterative least-squares procedure makes use, in effect, of another parameter in fitting the points—so it will often find a better fit as if there were a defect population even if none is present.

CDF ESTIMATION FROM RENEWAL DATA

The tracking of the performance of electronic components in actual field use is important to manufacturers for several reasons. This endeavor allows

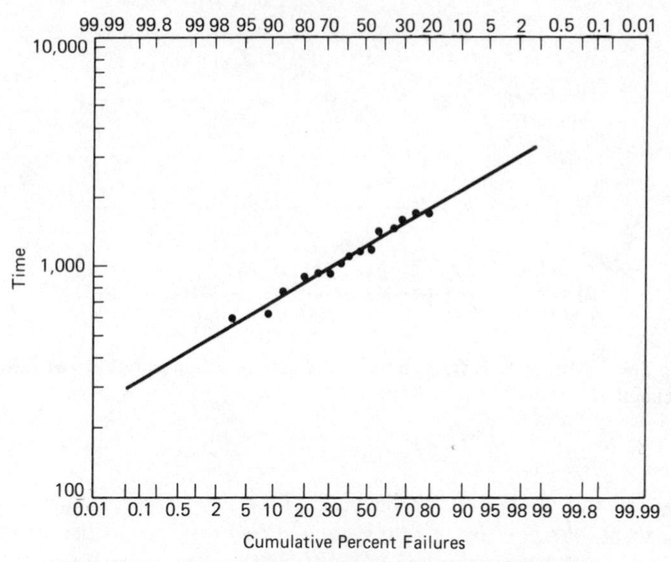

Figure 8.4. Data from Example 8.9 Plotted (Adjusted).

the producers to estimate component reliability as a function of operating time, check current product reliability models, and improve reliability projections for future technologies.

In previous chapters, we have shown how various distributions could be estimated from laboratory experiments. However, data from the field may not be available in a convenient form. In particular, let us consider a special type of data-generating procedure called a renewal process.

Suppose we have a system (e.g., computer) consisting of c components from the same parent population. The components fail independently with a lifetime distribution (CDF) given by $F(t)$. We wish to estimate $F(t)$, or equivalently the reliability function $R(t) = 1 - F(t)$, from system lifetime data. We will not assume a particular parametric form for the lifetime distribution, but will seek a nonparametric estimator of $F(t)$.

Upon failure of a component, a system is assumed to cease operation. However, the system is immediately restored to operation by replacement of the failed component with another component from the same population.

Moreover, we consider the case in which only the times to failure of the system, and not the failure times of the individual components, are recorded, a condition we refer to as unidentified replacement. This situation may arise because records are not kept of the site of each failed component and its replacements. For example, a drilling machine tool may have many bits that are individually replaced upon failure, but only the times of machine stoppage

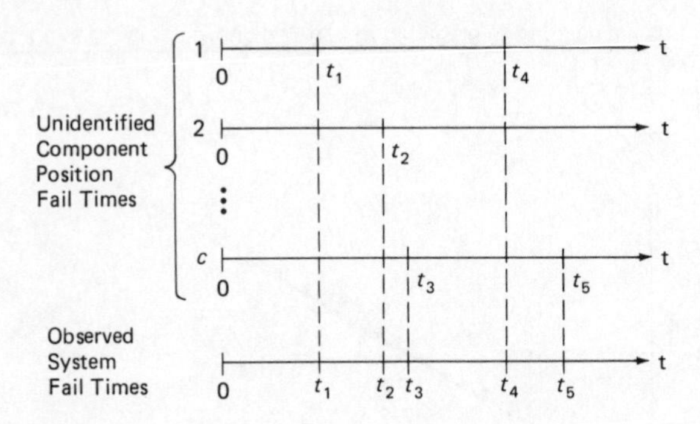

Figure 8.5. System of *c* Components Viewed as a Superposition of Renewal Processes.

are recorded and not the particular bit position where replacement occurred. Hence, except for a one component system, it becomes impossible to determine after the first fail whether any subsequent fails occurred on original components or on components that were replacements for original units.

Since we assume the original and replacement component lifetimes are a random sample from the component lifetime distribution $F(t)$, the sequence of interreplacement times for a single component position can be viewed as an ordinary renewal process described in Chapter 2. For a system, we then have a superposition of such renewal processes as shown in Figure 8.5. Given information on only the system failure times, how do we estimate the cumulative distribution function of the component lifetimes?

Our approach to estimating $F(t)$ relies on the fundamental renewal equation which relates $F(t)$ and the component renewal function $M(t)$ via a convolution integral. $M(t)$ is the expected number of component renewals or replacements made through time t. First, $M(t)$ is estimated from system fail times, and then $F(t)$ is estimated by numerical deconvolution of the renewal equation.

This topic has been extensively treated by Trindade and Haugh (1979–1980). They cover the multicensored situation of N systems having different operating hours. In addition, the statistical properties of various numerical deconvolution methods have been investigated. We shall employ one such method in this section.

A well-known relationship exists between the CDF $F(t)$ for component failure time and the renewal function $M(t)$. Called the fundamental renewal equation (Barlow and Proschan, 1975), the relation is

$$M(t) = F(t) + \int_0^t M(t-x)\, dF(x).$$

For our purposes, we write this equation in the equivalent form

$$F(t) = M(t) + \int_0^t M(x)\,dF(t-x).$$

While this equation looks foreboding, we will not be using it directly. Instead, we look for its use in some statistical applications.

An unbiased estimator of $M(t)$, for a single system of c components, is just

$$M(t) = \frac{n(t)}{c}.$$

where $n(t)$ is the number of renewals for all component positions by time t. Note the estimator $M(t)$ is a step function with jumps at the fail times. By numerically solving the fundamental renewal equation, Trindade and Haugh show that the following equations can be used to estimate the CDF at a given time t_1:

$$\begin{aligned} F(t_1) &= M(t_1) \\ F(t_2) &= M(t_2) - F(t_2 - t_1)M(t_1). \end{aligned}$$

We note at the second fail time that $F(t_2 - t_1)$ is required and $t_2 - t_1$ may be greater than, equal to, or less than t_1. By using the information that $F(t_1)$ is specified at t_1, we can develop a recursive approach such that

$$\begin{aligned} F(t_2 - t_1) &= 0, && \text{if } t_2 - t_1 < t_1 \\ &= F(t_1) && \text{if } t_2 - t_1 \geq t_1. \end{aligned}$$

Similarly,

$$F(t_3) = M(t_3) - F(t_3 - t_1)M(t_1) - F(t_3 - t_2)[M(t_2) - M(t_1)],$$

where

$$\begin{aligned} F(t_3 - t_j) &= 0 && \text{if } t_3 - t_j < t_1 \\ &= F(t_1) && \text{if } t_1 \leq t_3 - t_j < t_2 \\ &= F(t_2) && \text{if } t_2 \leq t_3 - t_j, \end{aligned}$$

for $j = 1, 2$. Thus, we see that differences, $\delta_{3j} = t_3 - t_j$, between the given fail time and each earlier time are compared to the actual fail times in order

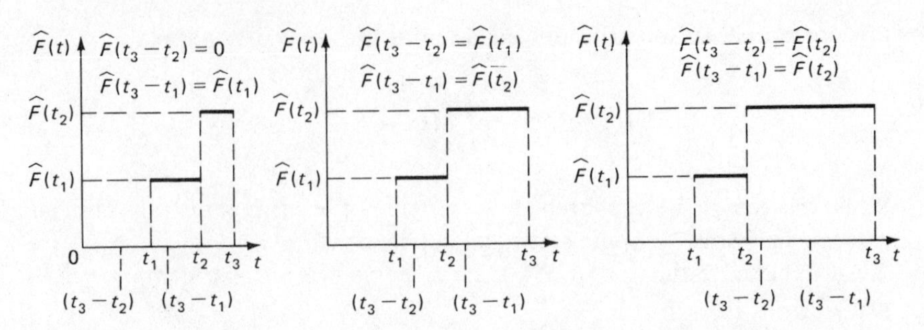

Figure 8.6. Possible Outcomes for Time Differences in Renewal Estimation.

to assign a proper $F(\delta_{3j})$ value from the previously calculated $F(t)$ estimates. See Figure 8.6 for a graphical representation of this procedure.

We will give the general equation for the CDF $F(t_k)$ at time t_k, but as we have shown above, the formulas can be written without using the detailed notation shown below. So for $k \geq 2$, we have

$$F(t_k) = M(t_k) - \sum_{j=1}^{k-1} F(t_k - t_j)[M(t_j) - M(t_{j-1})]$$

for $1 \leq j \leq k-1$, with $t_0 = 0$, $F(0) = M(0) = 0$, and denoting $\delta_{kj} = t_k - t_j$, we develop recursively,

$$\begin{aligned} F(\delta_{kj}) &= 0 && \text{if } \delta_{kj} < t_1 \\ &= F(t_i) && \text{if } t_i \leq \delta_{kj} < t_{i+1}; \quad 1 \leq i \leq k-2 \\ &= F(t_{k-1}) && \text{if } t_{k-1} \leq \delta_{kj}. \end{aligned}$$

The comparisons of time diferences to observed fail times to locate the proper recursive values for evaluating the CDF at time t can involve considerable computational effort as the number of fails increases. Trindade and Haugh present an equal interval method of deconvolution which avoids these comparisons and speeds up computations.

EXAMPLE 8.10 RENEWAL DATA CALCULATION OF CDF

Consider a system of 25 components. System failures were reported at 35, 79, 142, and 206 hr. Estimate the component CDF $F(t)$.

The renewal function estimates are:

$$M(35) = 1/25 = .04$$
$$M(79) = 2/25 = .08$$

$$M(142) = 3/25 = .12$$
$$M(206) = 4/25 = .16.$$

The CDF estimates are:

$$\begin{aligned}
F(35) &= M(35) = .04 \\
F(79) &= M(79) - F(44)M(35) = .08 - .04(.04) = .0784 \\
F(142) &= M(142) - F(107)M(35) - F(63)[M(79) - M(35)] \\
&= .12 - .0784(.04) - .04(.08 - .04) = .1153 \\
F(206) &= M(206) - F(171)M(35) - F(127)[M(79) - M(35)] - \\
&\quad F(64)[M(142) - M(79)] \\
&= .16 - .1153(.04) - .0784(.08 - .04) - .04(.12 - .08) \\
&= .1507.
\end{aligned}$$

SUMMARY

This chapter introduced the series and competing risk models, the escape discovery model, and the defect model as building blocks to system reliability. In conjunction with a general algorithm for component failure rates, we have a flexible set of tools to model and project reliability. If we have stress dependency incorporated into our general algorithm, we can also make trade-off analysis calculations concerning burn-in and test efficiency changes.

Methods of graphically spotting the possible presence of a reliability defective subpopulation were described, as was a procedure for estimating a CDF given field data of the renewal type.

We also saw how system redundancy provides enormous improvement in early life failure rates. As component costs go down, and high reliability becomes more essential, this design option will be used more and more.

9

Quality Control in Reliability: Applications of the Binomial Distribution

Manufacturers often conduct various tests on samples from individual lots in order to infer the expected level of reliability of product in customer applications. In addition, stressing of consecutive groups of samples from production may be employed to monitor the reliability of a process. Such results may be plotted on statistical control charts to verify that the manufacturing process is "under control." However, since such studies may be costly and time consuming—and often destructive to units—it is important that efficient sampling designs be selected to provide the necessary information while using the minimum quantities of product. This chapter covers the implementation of various types of sampling plans for attribute data, the associated risks, the operating characteristic curves, and the choice of minimum sample sizes. We discuss the applications of various discrete distributions, especially the binomial. Also, the calculation of confidence limits is treated. Finally, we look briefly at the application of statistical process control charting for reliability.

SAMPLING PLAN DISTRIBUTIONS

There are several important considerations an engineer must keep in mind when choosing appropriate sampling plans. First, one must establish the scope of the inference: are the results to be used to draw conclusions about an individual lot or about an ongoing process? If the former, then the size of the sample relative to the lot size must be considered. Obviously, if the sample size is a significant portion of the lot size, say over 10%, then there is more information about the quality of the remainder of the lot than there would be in sampling from a process capable of producing an infinite number of units. Indeed, different probability distributions are required to treat each case: the hypergeometric distribution applies when lot size must be considered; the binomial distribution holds for inference to a process. The binomial distribution has so many important applications it will be described in detail. However, in order to understand the simple derivation of the binomial distribution, we must digress for a moment and discuss permutations and combinations.

PERMUTATIONS AND COMBINATIONS

A permutation is an arrangement of objects in which order is important. For example, suppose we have three seating positions and three people: call them *A, B,* and *C.* We have three choices for placement in the first position: *A, B,* or *C.* Given occupancy of position one, there remains two possible choices for position two. That is, if *B* is in position one, only *A* and *C* are left to sit in position two. When both the first and the second positions are occupied, only one choice remains for the third seat. The fundamental principle states that if an operation has x possible outcomes and each of these outcomes can be followed by an operation consisting of y possible outcomes, then the total number of outcomes for the successive operations is xy. Extension to three or more operations is obvious.

So by this product rule for the seating arrangments, there are a total of $3 \times 2 \times 1 = 6$ possible orders. We list them below:

POSSIBLE ARRANGEMENTS OF THREE OBJECTS
ABC
ACB
BAC
BCA
CAB
CBA

These orders are called permutations. The descending product of numbers $n\ (n-1)\ (n-2) \cdots 1$ is denoted as $n!$, called "n factorial." By convention, $0! = 1$.

Suppose we have eight objects and we wish to determine the number of ordered arrangements we can form using four objects at a time. The first position may be occupied by any of eight objects, the second by any of the remaining seven objects, the third by any of the six, and the fourth by any of the five left. Thus, the total number of permutations is $8 \times 7 \times 6 \times 5 = 1680$. We will not list these, but the general formula for the number of permutations of n objects taken r at a time is

$$_nP_r = n(n-1)(n-2)\cdots(n-r+1).$$

Note there are r separate terms in the product above.

Suppose we have eight objects and we want to select four at a time, but we do not care about the order of the four selected. Such would be the case, for example, if we were forming teams consisting of four players each from eight possible choices. Such selections in which order is immaterial are called "combinations." To determine the number of combinations, one just divides the number of permutations calculated before by 4!, the number of permutations of four objects at a time. Thus, the number of combinations of eight objects taken four at a time is

$$\begin{aligned} _nC_r &= \frac{n(n-1)(n-2)\cdots(n-r+1)}{r!} \\ &= \frac{8 \times 7 \times 6 \times 5}{4 \times 3 \times 2 \times 1} = 70. \end{aligned}$$

In general, the expression for the number of combinations can multiplied by $1 = [(n-r)!/(n-r)!]$ to give a simpler appearing expression

$$\begin{aligned} _nC_r &= \frac{_nP_r}{r!} \\ &= \frac{n(n-1)(n-2)\cdots(n-r+1)}{r!} \\ &= \frac{n(n-1)(n-2)\cdots(n-r+1)(n-r)(n-r-1)\cdots 1}{[(n-r)(n-r-1)\cdots 1]r!} \\ &= \frac{n!}{(n-r)!r!} \\ &= \binom{n}{r}, \end{aligned}$$

where $\binom{n}{r}$ is the special symbol used to denote the number of combinations of n objects taken r at a time.

THE BINOMIAL DISTRIBUTION

The exponential, Weibull, normal, and lognormal distributions are examples of continuous distributions. However, data do not always occur in a continuous form. For example, the following questions involve discrete concepts:

1. If the probability of failure of a component at time t is 0.1, what is the probability of at least one failure in ten similar units placed on stress and run for time t?
2. What is the probability of exactly no failures or of one, two, ... or ten failures by time t?

Such questions usually can be handled via the binomial distribution. In fact, four conditions are necessary for the binomial distribution to apply, as follows:

1. Only two outcomes are possible (e.g., success or failure).
2. There is a fixed number (n) of trials.
3. There exists a fixed probability, p, of success from trial to trial.
4. The outcomes are independent from trial to trial.

In general, for reliability to work, the first two conditions are met; the third is assumed to hold, at least approximately; and the fourth is usually applicable. For example, an engineer stresses a fixed number of units (Condition 2). Each unit will either survive or fail the test (Condition 1), and the failure of one unit will not affect the probability of failure of the others (Condition 4). For similar reasons, we assume all units are equally likely to fail (Condition 3).

The general expression for the binomial distribution, giving the probability of exactly x failures in n trials with probability of success p per trial is

$$P(X = x) = \binom{n}{x} p^x (1 - p)^{n-x}.$$

Although we have not derived previous distributions, it is very illustrative of the application of the probability rules and combination concepts covered earlier in this text to develop the binomial distribution directly. Consider a component with "known probability of failure" by time t given by p. That is, $p = F(t)$, where $F(t)$ is the known CDF value. (Note that p can represent success or failure probabilities since only two outcomes are possible; one with probability p and the other with probability $q = 1 - p$.) Suppose we have n such units and we run them on stress to t hr. One possible result is that at the end of the experiment there are no failures. The probability of a single unit surviving is $1 - p$. The probability of all n independent units surviving is given by the product rule as

$$P(X=0)=(1-p)(1-p)\cdots(1-p) \qquad (n \text{ terms})$$
$$=(1-p)^n.$$

Consider the case of one failure occurring. A possible sequence (where F = failure and S = survival) might be

$$FSSS\cdots SSS,$$

that is, the unit in position one fails and the remaining $n - 1$ units survive. The probability of this specific sequence is

$$p(1-p)(1-p)\cdots(1-p)=p(1-p)^{n-1}.$$

However, we normally do not specify the order in which a unit failed—only that some unit did not survive. Thus, we are interested in how many ways any one part can fail out of n items, when order does not matter. The answer is given by the combination formula ${}_1C_n = n$. Since all sequences with one failure and $n - 1$ survivors have the same probability and there are n such mutually exclusive sequences, then by the union or addition rule for probabilities, the probability for exactly one fail is

$$P(X=1)=np(1-p)^{n-1}.$$

Similarly, the probability of a given sequence in which two failures occur and $n - 2$ survive is given by

$$p^2(1-p)^{n-2}.$$

Since there are ${}_nC_2 = n(n-1)/2$ different ways of choosing two failures from n units where order is immaterial, the probability of exactly two failures in n items is

$$P(X=2)=\frac{n(n-1)p^2}{2}(1-p)^{n-2}.$$

Continuing this way, we see in general that the probability of getting exactly x failures from n items on stress is

$$P(X=x)=\binom{n}{x}p^x(1-p)^{n-x}.$$

EXAMPLE 9.1 BINOMIAL CALCULATIONS

One hundred light bulbs will be stressed for 1000 hr. The probability of a bulb failing by 1000 hr has been determined to be .01 or 1% from previous experimental work. Assuming the bulbs on stress are from the same population,

a. What is the probability that all bulbs survive 1000 hr?
b. What is the probability of exactly one bulb failing?
c. What is the probability of at least one bulb failing?

SOLUTION

a. $(1-p)^n=(1.00-.01)^{100}=.99^{100}=.366.$

Therefore, approximately 1 chance out of 3 exists that no bulbs will fail by 1000 hr.

b. $np\ (1-p)^{n-1}=100(.01)\ (.99)^{99}=370.$
c. We could calculate the individual probabilities of 1, 2, 3, and so on, failures and add these together to get the answer. However, a much simpler procedure recognizes the fact that the probability of at least 1 failure equals 1 minus the probability of no failures. Thus,

$$P(X>0)=P(X\geq 1)=1-P(X=0)=1-.366=.634.$$

In roughly two out of three such experiments, we would expect at least 1 failure, but in only about one-third of the time will there be exactly 1 failure.

EXAMPLE 9.2 BINOMIAL PDF

Figure 9.1 shows a plot of the binomial probability density function for the case of 20 units sampled from a population with individual probability of surviving past 100 hr given by .2. The binomial distribution has an expected number of successes given by np and variance given by $np(1-p)$. Here, the expected number is $20 \times .2 = 4$ and the variance is $20 \times .2 \times .8 = 3.2$. Note, np corresponds to the peak in the probability distribution function.

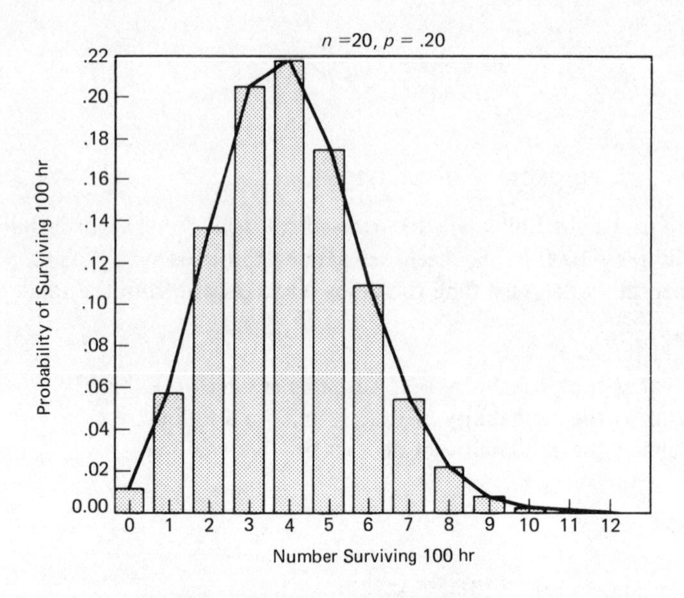

Figure 9.1. Binomial Distribution

NONPARAMETRIC ESTIMATES USED WITH BINOMIAL DISTRIBUTION

Suppose we have insufficient data to determine the underlying distribution for a component, but we know from nonparametric studies that about 95% of units survive 100 hr. We are planning a 100-hr mission requiring 20 such units to be operational. What is the probability of success for the mission, defined as no failures?

The nonparametric studies have provided the CDF estimate which we equate to the probability of failure *p*. Thus, the probability of no failures is

$$(1-p)^n = (1.00 - .95)^{20} = .359.$$

We realize that there is only about one chance in three, roughly, of having all components operational during the mission. We must go back to the drawing board to improve our reliability!

CONFIDENCE LIMITS FOR THE BINOMIAL DISTRIBUTION

Suppose we have a population of 50,000 integrated circuits and we wish to sample 100 units and stress for a particular mode of failure. At the end of the experiment, we wish to make a statement of inference about the probable range of percent defective in the population for this failure mode. That is, we will make an interval statement with a certain degree of confidence about the population, for example, we are 90% confident that the true percent defective is between 4.8 and 9.3%.

The simplest way to do this based on sample results is to refer to the classic charts for confidence limits for p in binomial sampling by Clopper and Pearson (1934), shown as Figures 9.2–9.5. Given the sample fraction defective y/n in an experiment, the confidence limits are obtained directly from the charts by the interesections of the curved n lines with the abscissa (x values) and then reading the ordinate (y values). For example, in our stress if we find 20 defective units out of 100, then $y/n = 20/100 = .20$. Referring to the chart for the 95% confidence level, we read the upper confidence limit from the upper curved $n = 100$ line intersecting with the abscissa $y/n = .20$ to be about .29. Similarly, the lower confidence limit appears to be about .13. Thus, we state with 95% confidence that the true population value lies somewhere within .13 to .29.

In using confidence limits we are effectively "tossing a horseshoe" at a population value. For 95% confidence, we expect that 19 out of 20 times

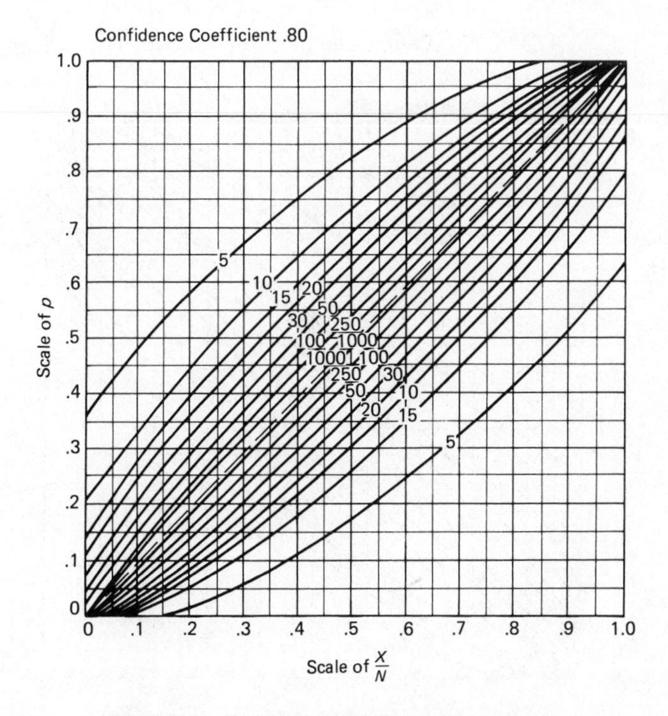

Figure 9.2. Confidence Belts for Proportions.
Figures 9.2–9.5 (Clopper and Pearson charts) reproduced by permission of *Biometrica* Trustees.

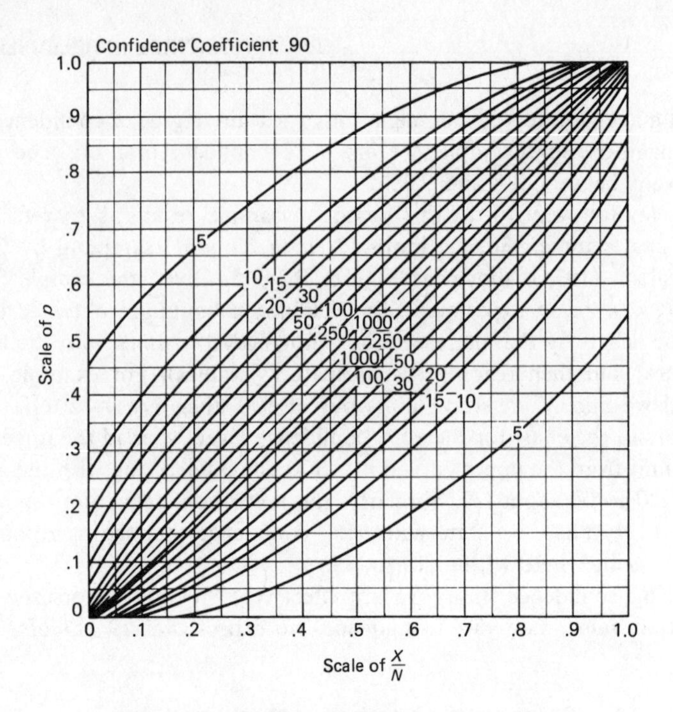

Figure 9.3. Confidence Belts for Proportions.

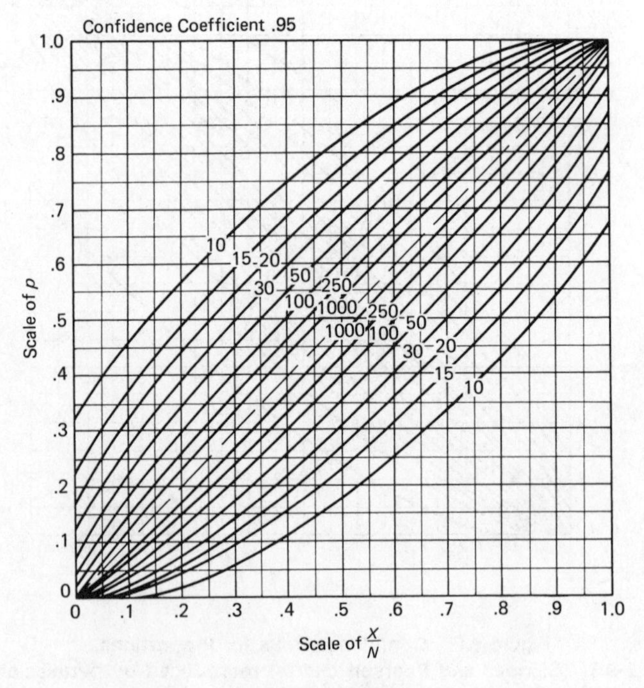

Figure 9.4. Confidence Belts for Proportions.

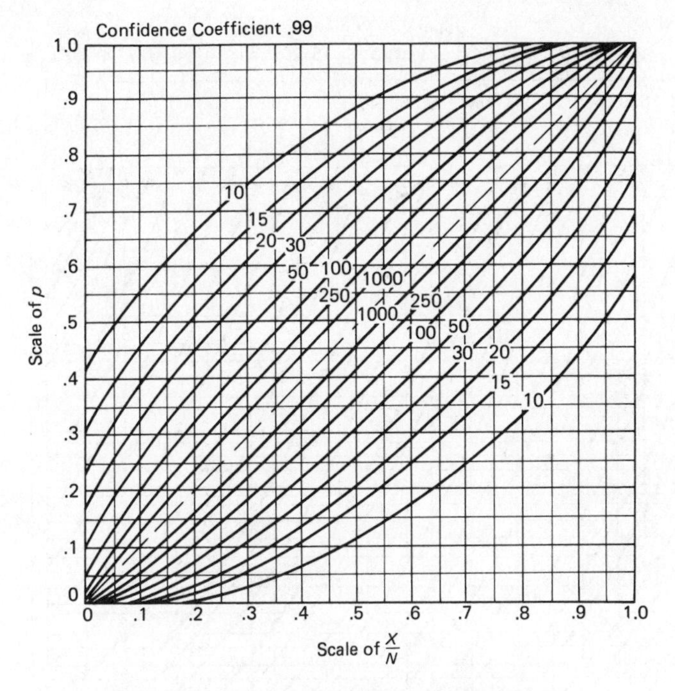

Figure 9.5. Confidence Belts for Proportions.

we should capture the population value within the limits given. To get greater confidence with the same sample size, we would need a larger horseshoe, that is, wider limits.

For small proportions, it becomes very difficult to accurately read the Clopper–Pearson charts. An improved chart was devised by R. Ament (1977). It is shown as Figure 9.6. To use this chart, let us assume two failures out of a sample of 200 units. We read the observed sample percent defective (1.0%) on the abscissa as before. However, we now assign to the ordinate the actual number of failures in the sample (here, 2). Note there are two sections to the ordinate: a numerator for the upper limit and a numerator for the lower limit. Finding the intersections of the ordinate values with the abscissa value provides, possibly with some visual interpolation, the upper and lower confidence limits for 95% confidence. For our case, we read .12% and 3.5% for the confidence limits. The very top line of the figure is used when zero failures in the sample are observed. In this case, we apply the sample size to the line marked "denominator" and read on the scale above the upper 95% confidence limit. For example, for zero failures out of 150 units, we have an upper 95% confidence limit of 2.4%.

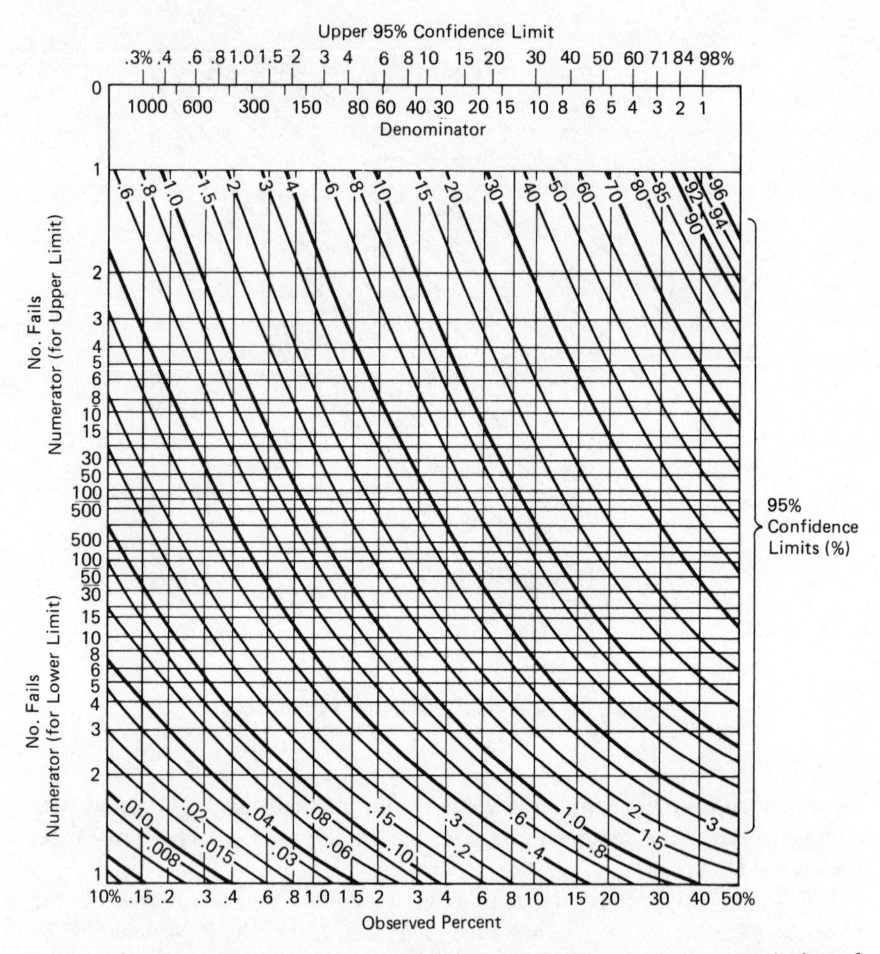

Figure 9.6. 95% Confidence Limits for Proportions. Reproduced by permission of the author, Richard P. Ament, commission on Professional and Hospital Activities)

HYPERGEOMETRIC AND POISSON DISTRIBUTIONS

A few words about two other important distributions are appropriate here The hypergeometric distribution is described by the following equation which gives the probability of getting $X = x$ rejects in a sample of size n drawn from a finite lot of size N containing a total of m rejects:

$$P(X=x)=\frac{\binom{n}{x}\binom{N-n}{m-x}}{\binom{N}{n}}, \qquad x=0,\ 1,\ 2,\ \ldots,\ n.$$

Note the denominator in the first expression is just the number of combinations of objects, taken n at a time, from N. Similarly the numerator is the total number of ways of getting x defectives in the sample of size n multiplied by the number of ways of getting $m - x$ defectives in the remaining group of size $N - n$. Because of the factorial terms, this formula is computationally difficult for large numbers. However, in the common situation where N is large relative to n, the binomial distribution gives an accurate approximation to the hypergeometric distribution.

EXAMPLE 9.3 HYPERGEOMETRIC DISTRIBUTION

The failure analysis lab has just received 10 units reported defective in an accelerated stress experiment. The electrical characteristics of the rejects are very similar, and the engineer does not have time to analyse all 10 units. He randomly chooses 4 parts for the failure autopsy. If there were actually 3 units with one mode of failure (say, type A) and 7 with another reason for failing (say, type B), what is the probability that none of the 3 examined units are type A?

SOLUTION

Use the hypergeometric distribution with $N = 10$, $n = 4$, $m = 3$; solve for $X = 0$:

$$P(X=0)=\frac{\binom{4}{0}\binom{10-4}{3-0}}{\binom{10}{4}}=\frac{1\times([6\times 5\times 4]/[3\times 2\times 1])}{(10\times 9\times 8\times 7)/(4\times 3\times 2\times 1)}=.095.$$

Hence, there is less than a 10% chance of not detecting both modes.

We also distinguish between a defect and a defective unit. A unit is defective if it has one or more defects, where a defect is defined as a nonconformance to specifications. Thus, the concept that a unit is defective is treated separately from the idea that there are a number of defects per unit or per area. The former case can be handled by the binomial (or hypergeometric) distribution. For calculations involving density (e.g., defects per wafer, failures per period, accidents per hour, etc.), the Poisson distribution is often employed. The equation for the Poisson distribution is

$$P(X=x)=\frac{\lambda^x e^{-\lambda}}{x!}, \qquad x=0,\ 1,\ 2,\ \ldots,$$

where λ = the average density. This simple formula can also be used to numerically approximate the binomial distribution, even though the methods of deriving the distributions are different. The approximation will be accurate when N is large compared to n, and the probability p that a unit is defective is small, say less than .10. To use the Poisson as an approximation to the binomial, we just use the average density, λ, set equal to the product $n \times p$.

EXAMPLE 9.4 POISSON DISTRIBUTION

A sample of 50 units is periodically drawn and tested. If the process producing these units generates a defect level of 5%, what is the probability of getting no failures in the sample tested?

SOLUTION

Use the Poisson distribution with $\lambda = 50 \times .05 = 2.5$ and calculate for $X = 0$:

$$P(X=0) = \frac{\lambda^x e^{-\lambda}}{x!} = e^{-2.5} = .082.$$

This result means that there is roughly only a 1 in 12 (i.e., the reciprocal of .082) chance of getting no failures.

The Poisson distribution also has a useful relationship to the exponential life distribution as follows:

A system with an exponential life distribution with parameter λ is repaired every time the system has a failure. The probability of having to make exactly k repairs in the time interval t is given by the Poisson distribution with parameter λt. In other words,

$$P(k \text{ repairs}) = (\lambda t)^k e^{-\lambda t}/k!.$$

As a confirmation of this relationship, consider that the probability of failure free operation in the time interval t is the probability of zero repairs, that is $k = 0$, or

$$P(\text{no repairs}) = e^{-\lambda t}.$$

As we showed in Chapter 1, the complementary event to no repairs is at least one repair, and this probability is given by

$$P(\text{at least one repair}) = 1 - e^{-\lambda t}.$$

This last expression is just the CDF $F(t)$ for the exponential distribution with parameter λ. So a discrete process (for example, counting the number of repairs) described by a Poisson distribution with parameter λt has a probability of failure in the time interval t described by the continuous exponential distribution with parameter λ.

TYPES OF SAMPLING

Let us consider a lot consisting of a very large number of packaged integrated circuits. Suppose we are concerned about a particular mode of failure. Assume that the population consisting of the entire lot would suffer a percent defective failure total of some value, say 1%, if the entire lot were stressed a certain length of time, say 100 hr, under normal operating conditions (voltages and temperature and humidity). However, we wish to test only a small sample for the same time under similar operating conditions to estimate the population percent defective and determine if the lot has acceptable reliability. How do we choose the appropriate sample size?

We will make a decision based on the results of looking at one sample of size n. This type of analysis is called "single sampling." It is also possible to perform double sampling wherein the results of the first sample drawn from a lot generate one of three possible outcomes: (1) accept the lot, (2) reject the lot, or (3) take an additional sample and combine the results to reach a final decision. In fact, multiple sampling plans are possible in which more than two draws from the population are performed before a decision is reached. For further discussion of multiple sampling plans, the reader should consult Burr (1976) or Schilling (1982). Also, MIL STD 105D (1963) contains actual single, double, and multiple sampling schemes for inspection by attributes.

In this chapter, only single-sample plans will be discussed. Since one either accepts or rejects a lot or process in single-sampling plans, the probability of acceptance plus the probability of rejection is equal to 1. Hence, rejection and acceptance probabilities will be used interchangeably.

RISKS

To fully understand the sampling process, one must comprehend the risks involved with the decision to either accept or reject a lot. A straightforward approach might be to establish an acceptable level or maximum level of allowed percent defective, say $Y\%$. Then, we could choose a convenient sample size for stressing, say 50 units, and establish the criteria for the maximum number of failures in the sample on stress so as to provide a specified probability of accepting or rejecting the lot. If we have more failures than the acceptance number (call it c), we will reject the lot; if we have c or fewer failures, we will accept the lot.

The acceptance number is based on the fact that a large lot with a fixed fraction defective will produce varying results for each sample drawn, even if the sample size is the same. We can calculate an acceptance number however, such that if the number of rejects in a sample exceeds it, then the probability is high that the lot has an excessive fraction defective. On the other hand, if the lot fraction defective is excessive, how likely are we to exceed this number in a given sample? Let's consider the risks associated with this simple plan.

First, we must realize that a matrix of possible correct and incorrect decisions exists, as shown in Table 9.1.

If the lot is truly less than $Y\%$ defective, and we make a decision based on the sample results (less than or equal to c failures from n units) to accept the entire lot, we have made a correct decision. If similarly, the lot is greater than $Y\%$ defective and we reject the lot based on the sample results, we again have made the correct decision. If the lot, however, is less than $Y\%$ defective and we end up rejecting the lot ($c + 1$ or more failures out of n units), then we have committed an error (called a Type I error), and the chance of reaching this wrong decision is referred to as an "α" (alpha) risk. Similarly, a Type II error at a "β" (beta) risk level occurs when we accept the lot, based on sample results, when the lot is actually greater than $Y\%$ defective.

Alternatively, the terms "producer's risk" and "consumer's risk" are often applied to the alpha and beta risks, respectively, since Type I error refers to the rejection of a good lot and Type II error designates the acceptance of a bad lot. Obviously it is costly to a producer to throw away or rescreen acceptable material due to Type I error. Similarly, the cost to the user of taking product with a high reject rate, because of Type II error, may be

Table 9.1 Matrix of Possible Choices.

		POPULATION VALUE (% DEFECTIVE)	
		$\leq Y\%$	$> Y\%$
DECISION ON LOT	Accept	Correct	Type II error
	Reject	Type I error	Correct

considerable if, for example, he has to replace or repair defective systems caused by bad components. Let us now consider the calculation of the probabilities of these possible choices.

OPERATING CHARACTERISTIC (OC) CURVE

If we are sampling from a finite lot, the distribution of the possible outcomes is correctly described by the hypergeometric model. However, when the sample size selected is small relative to the lot size, say less than 10% of the lot size, the binomial distribution is used as an excellent approximation to the hypergeometric distribution. We will assume this is the case, that is, the lot size is at least 10 times as great as the sample size.

Our first task is to calculate the probability of getting an acceptable lot based on c or less failures out of the n, say 50, units on stress. Let us assume that the lot percent defective is actually 2%. Let us further state that we wish to have an acceptance number c such that at least 95% of the time we accept the lot if the true percent defective is 2% or less; that is, our risk of Type I error is 5%. We now calculate the cumulative binomial probabilities for zero, 1, 2, and so on, up to c failures, for $n = 50$, $p = .02$, until the probability of getting c or less failures exceeds 95% or .95. Then 95% of the time, we will accept a lot with 2% defective if the decision to accept is based on c or less failures, because c or less failures are expected at least 95% of the time for a sample of size $n = 50$ drawn from a large lot with percent defective of 2% or less.

BINOMIAL CALCULATIONS

The formula for the cumulative binomial probability is

$$P(X \leq c) = \sum_{x=0}^{c} \binom{n}{x} p^x (1-p)^{n-x}.$$

So we require

$$\sum_{x=0}^{c} \binom{50}{x} (.02)^x (.98)^{50-x} \geq .95.$$

Note: This calculation can be performed simply on a scientific calculator with memory by using the following recursive formula to calculate successive binomial terms, starting with $(1-p)^n$:

$$P(X = x + 1) = \frac{(n-x)p\, P(X = x)}{(x+1)(1-p)}.$$

Thus,

$$P(X=0)=(1-.02)^{50}=.3642$$

$$P(X=1)=\frac{50(.02)\,P(X=0)}{1(.98)}=.3716$$

$$P(X=0)+P(X=1)=.7358$$

$$P(X=2)=\frac{49(.02)\,P(X=1)}{2(.98)}=.1858$$

$$P(X=0)+P(X=1)+P(X=2)=.9216$$

$$P(X=3)=\frac{48(.02)P(X=2)}{3(.98)}=.0607$$

$$P(X\leq 3)=P(X\leq 2)+P(X=3)=.9822.$$

At this point we stop, since the cumulative binomial probability, that is, $P(X \leq 3)$, is greater than .95. Alternatively, instead of calculating these terms individually, we could have used a table such as those in the *CRC Standard Mathematical Tables* (1965) which provide the cumulative binomial terms for various parameters. By either method, the desired c number is 3, that is, we accept a lot if 3 or less units fail out of the 50 sampled; reject the lot if 4 or more fail. Our risk of committing a Type I error will then be less than 5%.

Now what about the risk of a Type II error, that is, accepting a bad lot defined as over 2% defective? To calculate the risks associated with the Type II error it is not sufficient to state only one defect level; one needs to specify the various alternative percent defective values and perform each calculation separately. For example, one may be interested in the probability of getting 3 or less failures out of 50 units sampled (and thereby accepting the lot) if the true population value is 10, 15, 20, 30, or 40%, and so on. We have done such calculations and the numbers are shown as Table 9.2.

EXAMPLES OF OPERATING CHARACTERISTIC CURVES

This table may be better represented in the form of a graph, called the operating characteristic (OC) curve, which details the probability of accepting a lot based on c allowed rejects in a sample of size n. Such a curve is shown, for $n = 50$, $c = 3$, as Figure 9.7.

The operating characteristic curve provides the total picture of the sampling plan. We note, for example, that if a lot having 7% defective units is presented for sampling, there is about a 50–50 chance of its being accepted; we call this percent defective the "point of indifference." Also, we have about a

Table 9.2. Probability of 3 or Less Failures in Sample of Size $n = 50$ for Various Lot Percent Defective Values

LOT PERCENT DEFECTIVE, p	PROBABILITY, $P(\leq 3)$
.01	.9984
.02	.9822
.03	.9372
.05	.7604
.07	.5327
.10	.2503
.13	.0958
.15	.0461

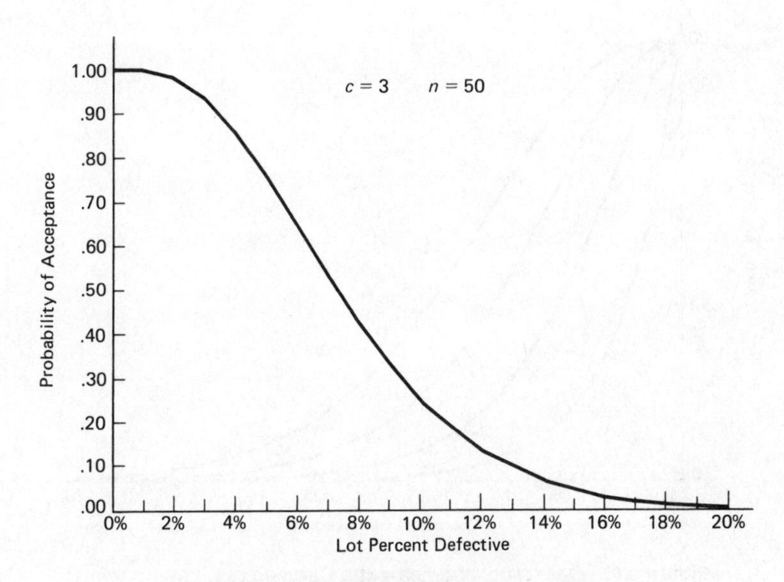

Figure 9.7. Operating Characteristic Curve.

10% chance of accepting a lot offered at around 12.5% defective, or over six times the 2% value we targeted in setting up this sampling plan. Stated another way, the consumer's risk is 10% that a lot at 12.5% will have three or less failures in a sample of 50 and thus be accepted.

How can the Type II or consumer's risk be reduced? Obviously one can decrease the acceptance number to $c = 2$, 1, or even zero, but what would that change mean to the OC curve? Figure 9.8 shows the OC curve for various acceptance numbers with the sample size fixed at 50. We see that the probability of acceptance has been reduced at all incoming percent defective levels. At the target value of 2% the probability of acceptance goes down from .98 to .92 to .74 to .36% as we change c from 3 to 2 to 1 to 0, respectively. Thus, we have greatly increased the probability of rejecting an acceptable lot, that is, the producer's risk, by lowering the acceptance number while keeping the sample size fixed.

Similarly, if only the sample size increases and the acceptance number c is held constant, then again the consumer's risk can be reduced but only at the expense of the producer's risk. Figure 9.9 shows the OC curve for samples of size $n = 50$, 100, 300, and 500 and acceptance number $c = 3$. The affect of increasing the sample size while the acceptance number remains fixed is to significantly reduce the probability of a lot acceptance at all percent defective values. Thus, the consumer's risk is dramatically driven lower, but the

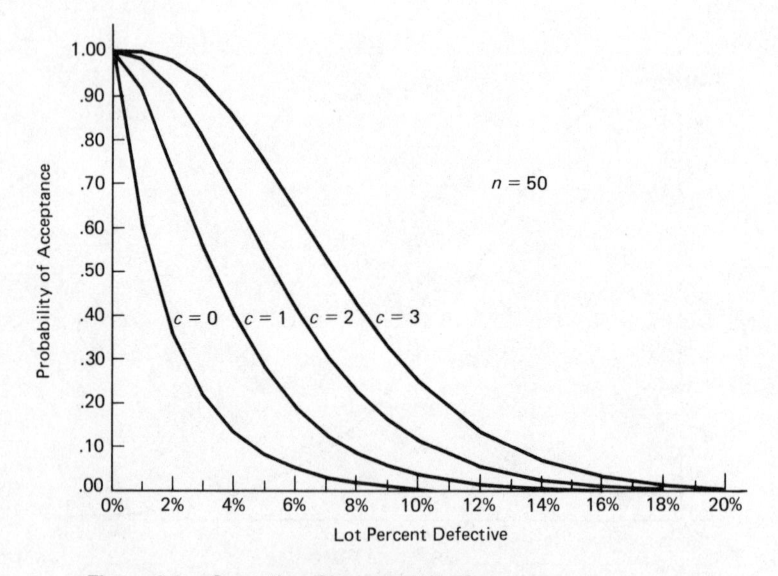

Figure 9.8. Operating Characteristic Curves. For Various Accept (c) Numbers. $n = 50$

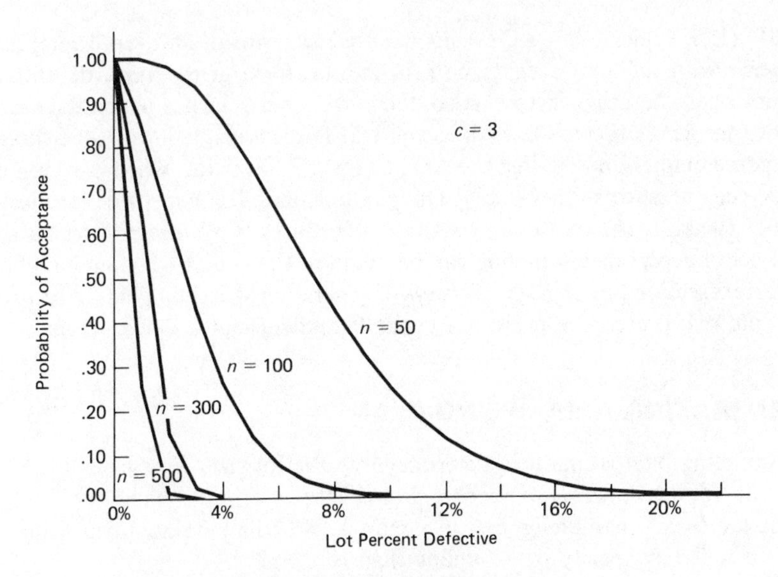

Figure 9.9. Operating Characteristic Curves. Fixed Acceptance No. Various *SS*

producer's risk becomes very high. For example, the probability of accepting a lot at 2% defective, allowing three rejects in sample of size $n = 50$ is about 98%; hence, the Type I risk is only 2%. However, a sample of size 100, 300, and 500 will have a probabilities of acceptance of 86, 15, and 1%, respectively, for the same acceptance number $c = 3$. So the producer's risk is correspondingly 14, 85, and 99%. However, for the consumer's risk held at 10%, the corresponding percent defective values are approximately 13, 6.5, 2.2, and 1.1% for samples of size 50, 100, 300, and 500, respectively.

In fact the only way to decrease both Type I and II errors simultaneously is to increase the sample size drawn and adjust the acceptance number. What we would like to do, indeed, is to specify a high probability of acceptance, say 95%, at a desirable or acceptable quality level and simultaneously require a high probability of rejection, say 90%, at an undesirable or rejectable quality level. For individual lots, it is common to call the percent defective value at the 95% acceptance probability the AQL for acceptable quality level; similarly, the 10% acceptance probability is referred to as the LTPD or lot tolerance percent defective or even sometimes as the RQL for rejectable quality level.

Note, however, that the term AQL is often applied in the literature in the context of an expected quality level without being tied to a given probability point, such as 95%. Furthermore, military documents, such as MIL STD

105D (1963) refer to an AQL value assured in terms of an overall sampling scheme with different acceptance probabilities for a given AQL, depending on lot sizes and other factors. Also, there are other schemes which are based on a specific characteristic such as the LTPD for a single lot or the average outgoing quality limit (called the AOQL) for a series of lots when it is feasible to screen out defects. The book by Dodge and Romig (1959) provides extensive tables for both the LTPD and AOQL procedures. Full information on the subject of acceptance sampling can be found in the book by Schilling (1982). The next section presents several ways of generating sampling plans to provide specific risk protection at given acceptable and rejectable defect levels.

GENERATING A SAMPLING PLAN

A sampling plan is uniquely determined by the lot size, N, the sample size, n, and the acceptance number, c. For a large lot or for sampling from a process, the N parameter can be ignored. We shall do so here. Thus, n and c uniquely specify the sampling plan.

The simplest way to get n and c for a given set of percent defectives and matching α,β risks is to find a computer program that does the calculations. For example, IBM has a software package called STATLIB written in APL by Jack Prins (1984) for the IBM PC or XT or equivalent. The format is simple: one types an expression like

.025 .12 SAMPLINGPLAN .05 .10

and the computer responds with the sample size and acceptance number necessary in order to have the AQL = 2.5% and the LTPD = 12%. The user must select the two levels of risk and the corresponding percent defectives. Also, the program lists the OC curve and provides a plot if requested. An example of the output is shown as Figure 9.10.

Another simple procedure is to employ the nomograph shown as Figure 9.11. To use the graph, two vertical scales are provided: the scale on the left refers to the fraction defective values (the x axis on the OC curve); the right scale designates the fractional probability of acceptance (the y axis on the OC curve). Draw two lines: for example, the first may extend from the AQL value on the left vertical scale to the matching .95 probability of acceptance on the right; the second line may go from the LTPD value on the left to the matching .10 probability on the right. The intersection of the two lines in the center grid determines the sample size n and the acceptance number c (called number of occurrences here). For the case where the AQL = 2% and the LTPD = 8%, we get $n = 98$ and $c = 4$.

```
      TRYPLAN
FIND A SAMPLING PLAN WITH NO MORE THAN 10 PERCENT CHANCE OF
ACCEPTING A LOT WHOSE TRUE PERCENT DEFECTIVES IS 12 OR WORSE,
AND 5 PERCENT CHANCE OF REJECTING WHEN THE TRUE PERCENT DEFECTIVES
IS 2.5 OR BETTER, FOR EXAMPLE:
     P1 = .025

     P2 = .12

     ALPHA,THE PRODUCER'S RISK = .05
     BETA, THE CONSUMER'S RISK = .10

THE PROGRAM IS ABOUT TO EXECUTE:
     .025 .12 SAMPLINGPLAN .05 .10

THE SAMPLE SIZE IS          55
THE ACCEPTANCE NUMBER IS    3
DO YOU WANT TABULAR OUTPUT?  Y/N
Y
PROBABILITY  PROPORTION
    OF           OF
ACCEPTANCE    DEFECTIVES
   .9977        .0100
   .9622        .0230
   .8641        .0360
   .7170        .0490
   .5533        .0620
   .4011        .0750
   .2756        .0880
   .1807        .1010
   .1138        .1140
   .0691        .1270
   .0406        .1400
   .0231        .1530
   .0128        .1660
   .0069        .1790
   .0036        .1920
   .0019        .2050
   .0009        .2180
   .0005        .2310
   .0002        .2440
   .0001        .2570
   .0000        .2700
DO YOU WANT AN OC CURVE?  Y/N
Y
```

```
1.0|
0.9|
0.8|
0.7|
0.6|
0.5|
0.4|
0.3|
0.2|
0.1|
0.0|
0.00   0.05   0.10   0.15   0.20   0.25   0.30
THE AOQL =  .0351
```

Figure 9.10. Computer Generated Sampling Plan

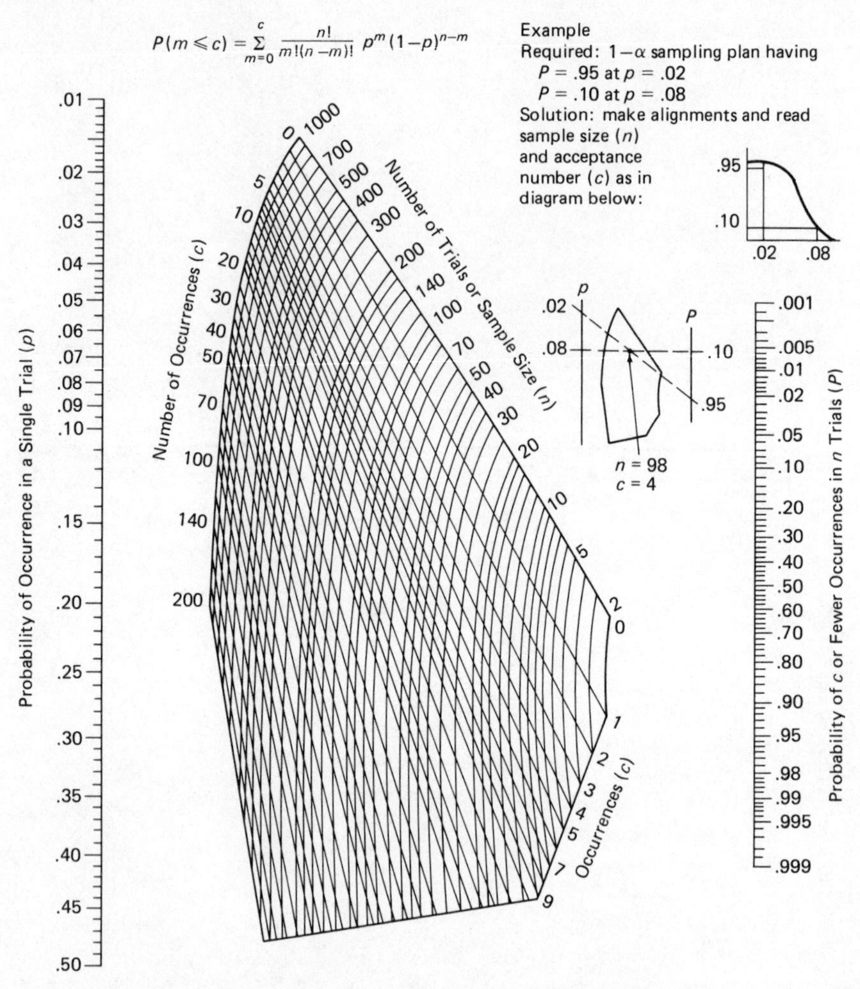

Note: If p is less .01, set kp on the p scale and n/k on the n scale, where $k = .01/p$, rounded upward convently.

Figure 9.11. Nomograph of the Cumulative Binomial Distribution. (Reprinted with permission from AT&T Technologies, Inc.)

Guenther (1974) presents a very simple iterative procedure, using the percentiles $\chi^2_{df;\,100\,\times\,\text{area}}$ of the chi-square distribution in the Appendix table. The equation for determining the minimum sample size n and accompanying acceptance number c is

$$.5 \times \left[\chi^2_{2c+2;\,100\times(1-\beta)} \times \left(\frac{1}{p_2} - .5 \right) + c \right] \le n \le .5 \times \left[\chi^2_{2c+2;\,100\times\alpha} \times \left(\frac{1}{p_1} - .5 \right) + c \right].$$

For a given set of conditions (p_1, p_2, α, β), one simply tries a c value, using the chi-square percentiles for $2c + 2$ degrees of freedom and the respective probabilities, and determines if the inequalities for the sample size n are mathematically satisfied. If so, then the minimum n is chosen. If not, then the c value is incremented by 1, until the interval for n contains at least one integer. For example, let us find a sampling plan to accept 95% of the time product at 2% defective and reject 90% of the time product at 8% defective. If we try $c = 0$, 1, 2, or 3 in the above formula, we find that no solutions exist. To illustrate for $c = 3$, we get

$$.5 \times \left[\chi^2_{8;\,90} \left(\frac{1}{.08} - .5 \right) + 3 \right] \le n \le .5 \times \left[\chi^2_{8;\,5} \left(\frac{1}{.02} - .5 \right) + 3 \right]$$
$$.5 \times [13.36(12) + 3] \le n \le .5 \times [2.73(49.50) + 3]$$
$$81.66 \le n \le 69.07$$

and hence no solutions exists. However, for $c = 4$, we discover

$$.5 \times [\chi^2_{10;90}(12) + 4] \le n \le .5 \times [\chi^2_{10;\,5}(49.50) + 4]$$
$$.5 \times [15.99(12) + 4] \le n \le .5 \times [3.94(49.50) + 4]$$
$$97.94 \le n \le 99.52$$

and hence the minimum sample size is 98.

Alternatively, one may use a specially prepared table such as the one from MIL-S-19500G (1965), shown as Table 9.3, for LTPD sampling plans. The use of this table is quite simple. The LTPD is listed across the top as a percentage. The acceptance number c is the vertical left column. The intersection of a specified LTPD and acceptance number is the minimum sample size required. Below each sample size n in parentheses is the AQL (that is, the defect level accepted 95% of the time) for that (n,c) pair. For example, to provide an LTPD of 2% with an acceptance number of 2, one would need at least 195 units to test. The AQL for this sampling plan of n = 195 and $c = 2$ is .18%.

Table 9.3. *LTPD sampling plans*[1,2]

Minimum size of sample to be tested to assure, with a 90 percent confidence, that a lot having percent-defective equal to the specified LTDP will not be accepted (single sample).

Max. Percent Defective (LTPD) or λ	50	30	20	15	10	7	5	3	2	1.5	1	0.7	0.5	0.3	0.2	0.15	0.1
Acceptance Number (c) (r = c + 1)	Minimum Sample Sizes (For device-hours required for life test, multiply by 1000)																
0	5 (1.03)	8 (0.64)	11 (0.46)	15 (0.34)	22 (0.23)	32 (0.16)	45 (0.11)	76 (0.07)	116 (0.04)	153 (0.03)	231 (0.02)	328 (0.02)	461 (0.01)	767 (0.007)	1152 (0.005)	1534 (0.003)	2303 (0.002)
1	8 (4.4)	13 (2.7)	18 (2.0)	25 (1.4)	38 (0.94)	55 (0.65)	77 (0.46)	129 (0.28)	195 (0.18)	258 (0.14)	390 (0.09)	555 (0.06)	778 (0.045)	1296 (0.027)	1946 (0.018)	2592 (0.013)	3891 (0.009)
2	11 (7.4)	18 (4.5)	25 (3.4)	34 (2.24)	52 (1.6)	75 (1.1)	105 (0.78)	176 (0.47)	266 (0.31)	354 (0.23)	533 (0.15)	759 (0.11)	1065 (0.080)	1773 (0.045)	2662 (0.031)	3547 (0.022)	5323 (0.015)
3	13 (10.5)	22 (6.2)	32 (4.4)	43 (3.2)	65 (2.1)	94 (1.5)	132 (1.0)	221 (0.62)	333 (0.41)	444 (0.31)	668 (0.20)	953 (0.14)	1337 (0.10)	2226 (0.062)	3341 (0.041)	4452 (0.031)	6681 (0.018)
4	16 (12.3)	27 (7.3)	38 (5.3)	52 (3.9)	78 (2.6)	113 (1.8)	158 (1.3)	265 (0.75)	398 (0.50)	531 (0.37)	798 (0.25)	1140 (0.17)	1599 (0.12)	2663 (0.074)	3997 (0.049)	5327 (0.037)	7994 (0.025)
5	19 (13.8)	31 (8.4)	45 (6.0)	60 (4.4)	91 (2.9)	131 (2.0)	184 (1.4)	308 (0.85)	462 (0.57)	617 (0.42)	927 (0.28)	1323 (0.20)	1855 (0.14)	3090 (0.085)	4638 (0.056)	6181 (0.042)	9275 (0.028)
6	21 (15.6)	35 (9.4)	51 (6.6)	68 (4.9)	104 (3.2)	149 (2.2)	209 (1.6)	349 (0.94)	528 (0.62)	700 (0.47)	1054 (0.31)	1503 (0.22)	2107 (0.155)	3509 (0.093)	5267 (0.062)	7019 (0.047)	10533 (0.031)
7	24 (16.6)	39 (10.2)	57 (7.2)	77 (5.3)	116 (3.5)	166 (2.4)	234 (1.7)	390 (1.0)	589 (0.67)	783 (0.51)	1178 (0.34)	1680 (0.24)	2355 (0.17)	3922 (0.101)	5886 (0.067)	7845 (0.051)	11771 (0.034)
8	26 (18.1)	43 (10.9)	63 (7.7)	85 (5.6)	128 (3.7)	184 (2.6)	258 (1.8)	431 (1.1)	648 (0.72)	864 (0.54)	1300 (0.36)	1854 (0.25)	2599 (0.18)	4329 (0.108)	6498 (0.072)	8660 (0.054)	12995 (0.036)
9	28 (19.4)	47 (11.5)	69 (8.1)	93 (6.0)	140 (3.9)	201 (2.7)	282 (1.9)	471 (1.2)	709 (0.77)	945 (0.58)	1421 (0.38)	2027 (0.27)	2842 (0.19)	4733 (0.114)	7103 (0.077)	9468 (0.057)	14206 (0.038)
10	31 (19.9)	51 (12.1)	75 (8.4)	100 (6.3)	152 (4.1)	218 (2.9)	306 (2.0)	511 (1.2)	770 (0.80)	1025 (0.60)	1541 (0.40)	2199 (0.28)	3082 (0.20)	5133 (0.120)	7704 (0.080)	10268 (0.060)	15407 (0.040)
11	33 (21.0)	54 (12.8)	83 (8.3)	111 (6.2)	166 (4.2)	238 (2.9)	332 (2.1)	555 (1.2)	832 (0.83)	1109 (0.62)	1664 (0.42)	2378 (0.29)	3323 (0.21)	5546 (0.12)	8319 (0.083)	11092 (0.062)	16638 (0.042)
12	36 (21.4)	59 (13.0)	89 (8.6)	119 (6.5)	178 (4.3)	254 (3.0)	356 (2.2)	594 (1.3)	890 (0.86)	1187 (0.65)	1781 (0.43)	2544 (0.3)	3562 (0.22)	5936 (0.13)	8904 (0.086)	11872 (0.065)	17808 (0.043)
13	38 (22.3)	63 (13.4)	95 (8.9)	126 (6.7)	190 (4.5)	271 (3.1)	379 (2.26)	632 (1.3)	948 (0.89)	1264 (0.67)	1896 (0.44)	2709 (0.31)	3793 (0.22)	6321 (0.134)	9482 (0.089)	12643 (0.067)	18964 (0.045)
14	40 (23.1)	67 (13.8)	101 (9.2)	134 (6.9)	201 (4.6)	288 (3.2)	403 (2.3)	672 (1.4)	1007 (0.92)	1343 (0.69)	2015 (0.46)	2878 (0.32)	4029 (0.23)	6716 (0.138)	10073 (0.092)	13431 (0.069)	20146 (0.046)
15	43 (23.3)	71 (14.1)	107 (9.4)	142 (7.1)	213 (4.7)	305 (3.3)	426 (2.36)	711 (1.41)	1066 (0.94)	1422 (0.71)	2133 (0.47)	3046 (0.33)	4265 (0.235)	7108 (0.141)	10662 (0.094)	14216 (0.070)	21324 (0.047)
16	45 (24.1)	74 (14.6)	112 (9.7)	150 (7.2)	225 (4.8)	321 (3.37)	450 (2.41)	750 (1.44)	1124 (0.96)	1499 (0.72)	2249 (0.48)	3212 (0.337)	4497 (0.241)	7496 (0.144)	11244 (0.096)	14992 (0.072)	22487 (0.048)
17	47 (24.7)	79 (14.7)	118 (9.86)	158 (7.36)	236 (4.93)	338 (3.44)	473 (2.46)	788 (1.48)	1182 (0.98)	1576 (0.74)	2364 (0.49)	3377 (0.344)	4728 (0.246)	7880 (0.148)	11819 (0.098)	15759 (0.074)	23639 (0.049)
18	50 (24.9)	83 (15.0)	124 (10.0)	165 (7.54)	248 (5.02)	354 (3.51)	496 (2.51)	826 (1.51)	1239 (1.0)	1652 (0.75)	2478 (0.50)	3540 (0.351)	4956 (0.251)	8260 (0.151)	12390 (0.100)	16520 (0.075)	24780 (0.050)
19	52 (25.5)	86 (15.4)	130 (10.2)	173 (7.76)	259 (5.12)	370 (3.58)	518 (2.56)	864 (1.53)	1296 (1.02)	1728 (0.77)	2591 (0.52)	3702 (0.358)	5183 (0.256)	8638 (0.153)	12957 (0.102)	17276 (0.077)	25914 (0.051)
20	54 (26.1)	90 (15.6)	135 (10.4)	180 (7.82)	271 (5.19)	386 (3.65)	541 (2.60)	902 (1.56)	1353 (1.04)	1803 (0.78)	2705 (0.52)	3864 (0.364)	5410 (0.260)	9017 (0.156)	13526 (0.104)	18034 (0.078)	27051 (0.052)
25	65 (27.0)	109 (16.1)	163 (10.8)	217 (8.08)	326 (5.38)	466 (3.76)	652 (2.69)	1086 (1.61)	1629 (1.08)	2173 (0.807)	3259 (0.538)	4656 (0.376)	6518 (0.269)	10863 (0.161)	16295 (0.108)	21726 (0.081)	32589 (0.054)

[1] Sample sizes are based upon the Poisson exponential binomial limit.

[2] The minimum quality (approximate AQL) required to accept on the average 19 of 20 lots is shown in parenthesis for information only.

FURTHER GRAPHICAL TECHNIQUES FOR OBTAINING AN OC CURVE

While it is possible to search through the literature to find specific OC curves or sampling plans, it is very useful to have a simple graphical procedure that allows the engineer to quickly calculate an OC curve, assess sample requirements, or check risks for many different situations. It is for these reasons that we have developed Figures 9.12–9.15. Table 9.4 contains the actual values used to plot these curves. The description of the use of these figures will reveal their flexibility and convenience in handling many sampling problems.

There are four separate figures provided, one for each acceptance number $c = 0$, 1, 2, and 3. It is our experience in reliability work that because of the cost of destructive testing of units, sampling plans are generally restricted to low acceptance numbers. Hence, these figures should suffice for most situations. The abscissa (x axis) is the sample size n required; the scale is from 10 to 100,000 units. The ordinate (y axis) is the population percent defective p; the lowest scale is from .001 (10 PPM) to 100%. The lines running diagonally across the graph from upper left to lower right represent the constant

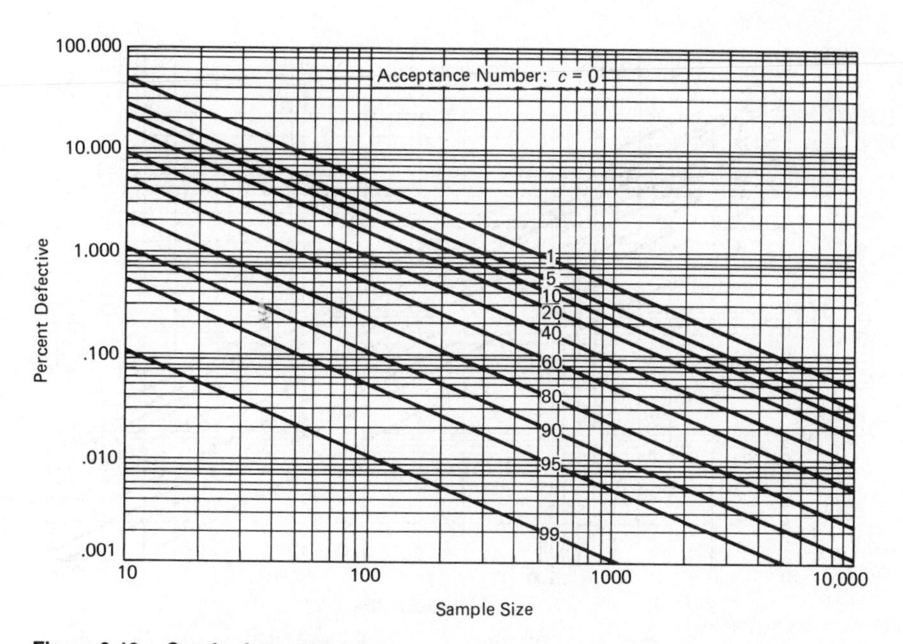

Figure 9.12. Graph of percent defective versus sample size. Diagonal lines represent various acceptance probabilities.

Acceptance Number: $c = 0$

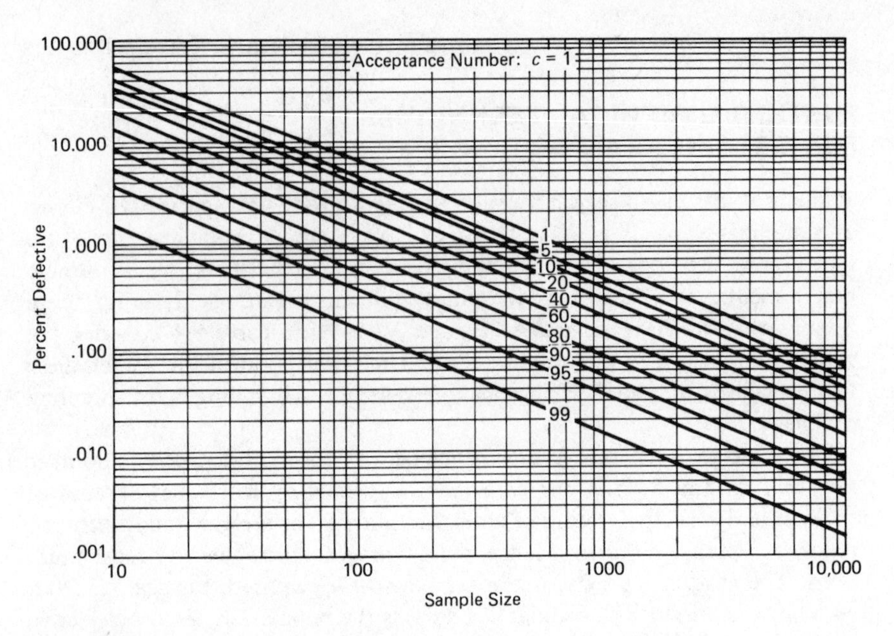

Figure 9.13. Graph of percent defective versus sample size. Diagonal lines represent various acceptance probabilities.

Acceptance Number: $c = 1$

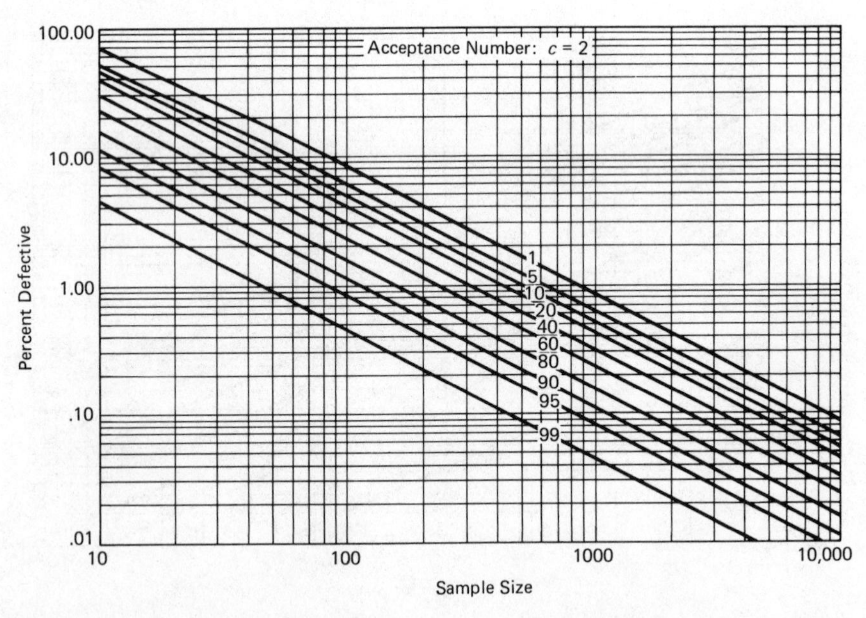

Figure 9.14. Graph of percent defective versus sample size. Diagonal lines represent various acceptance probabilities.

Acceptance Number: $c = 2$

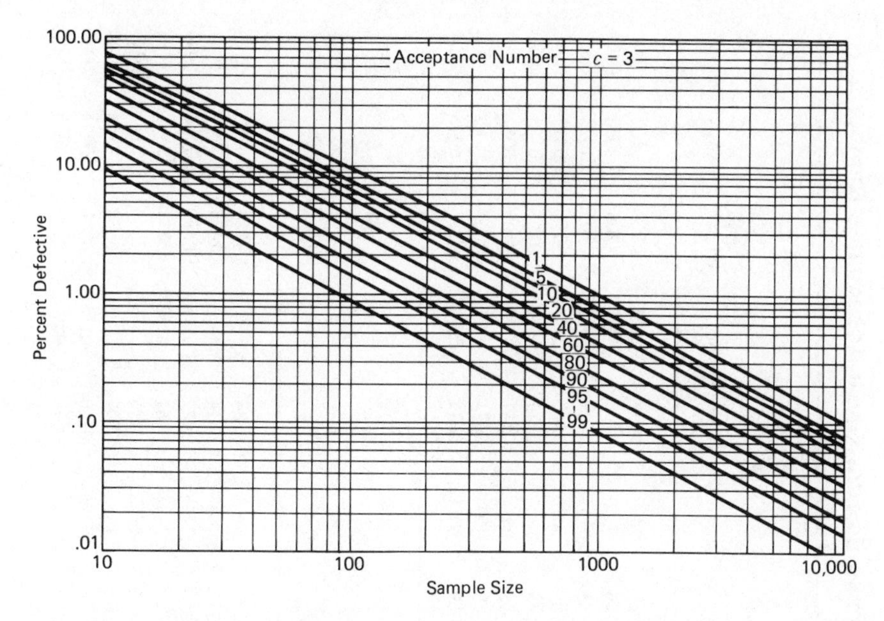

Figure 9.15. Graph of percent defective versus sample size. Diagonal lines represent various acceptance probabilities.
Acceptance Number: $c = 3$

probability of acceptance for the intersecting (n,p) points for the specific acceptance number c; the probability range runs from 1 to 99%. The 95 and 10% lines, corresponding to the AQL and LTPD specifications, respectively, are emphasized. These curves are calculated directly from the binomial distribution, and thus, may be applied in the situations where the sample size is small relative to the lot size or where we are sampling from a continuous process.

To generate an OC curve for a given sample size and acceptance number, one simply selects the figure corresponding to the acceptance number and finds the sample size on the x axis. Then, the percent defective p values are read off the vertical axis matching the probability of acceptance values 99%, 95%(AQL), 90%, 80%, 60%, 40%, 20%, 10%(LTPD), 5%, and 1%. Alternatively, Table 9.4 may be used. For example, for $n = 100$ and $c = 1$, the respective p values—for the horizontal axis of the OC curve we will form—are .149%, .357%, .533%, .825%, 1.37%, 2.01%, 2.96%, 3.83%, 4.66%, and 6.46%. From these values we generate the curve shown as Figure 9.16.

When it comes to comparing risk levels of various sampling plans or deciding on alternative sampling plans while holding a characteristic such as the

Table 9.4. Tables of Percent Defective Versus Sample Size for Given Acceptance Number.

$c = 0$

SAMPLE SIZE	PROBABILITY OF ACCEPTANCE 1%	5%	10%	20%	30%	40%	50%	60%	70%	80%	90%	95%	99%
10	37.0000	25.8870	20.5680	14.8670	11.3440	8.7557	6.6968	4.9800	3.5039	2.2068	1.0490	.5117	.1005
25	16.8240	11.2929	8.7990	6.2350	4.7018	3.5989	2.7346	2.0226	1.4166	.8887	.4206	.2050	.0402
50	8.7990	5.8156	4.5008	3.1677	2.3792	1.8159	1.3768	1.0165	.7109	.4453	.2105	.1025	.0201
100	4.6000	2.9514	2.2763	1.5966	1.1968	.9122	.6908	.5095	.3560	.2229	.1054	.0513	.0101
500	.9169	.5974	.4595	.3214	.2406	.1831	.1385	.1021	.0713	.0446	.0211	.0103	.0020
1,000	.4595	.2991	.2300	.1608	.1203	.0916	.0693	.0511	.0357	.0223	.0105	.0051	.0010
5,000	.0921	.0599	.0461	.0322	.0241	.0183	.0139	.0102	.0071	.0045	.0021	.0010	.0002
10,000	.0461	.0300	.0230	.0161	.0120	.0092	.0069	.0051	.0036	.0022	.0011	.0005	.0001

$c = 1$

SAMPLE SIZE	PROBABILITY OF ACCEPTANCE 1%	5%	10%	20%	30%	40%	50%	60%	70%	80%	90%	95%	99%
10	50.4400	39.4170	33.6850	27.0989	22.6946	19.2150	16.2270	13.5134	10.9284	8.3260	5.4529	3.6780	1.5539
25	23.7490	17.6130	14.6868	11.5089	9.4799	7.9256	6.6240	5.4648	4.3814	3.3098	2.1478	1.4404	.6047
50	12.5530	9.1399	7.5590	5.8704	4.8088	4.0035	3.3350	2.7426	2.1927	1.6518	1.0687	0.7154	.2997
100	6.4550	4.6560	3.8340	2.9646	2.4218	2.0130	1.6727	1.3738	1.0969	.8252	.5331	.3566	.1492
500	1.3300	.9453	.7757	.5977	.4872	.4041	.3355	.2752	.2195	.1649	.1064	.0711	.0297
1,000	.6620	.4735	.3885	.2991	.2438	.2021	.1678	.1376	.1097	.0825	.0532	.0356	.0149
5,000	.1327	.0949	.0778	.0599	.0488	.0404	.0336	.0275	.0220	.0165	.0106	.0071	.0030
10,000	.0664	.0474	.0389	.0300	.0244	.0202	.0168	.0138	.0110	.0082	.0053	.0036	.0015

$c = 2$

PROBABILITY OF ACCEPTANCE

SAMPLE SIZE	1%	5%	10%	20%	30%	40%	50%	60%	70%	80%	90%	95%	99%
10	61.1800	50.70000	44.9700	38.0938	33.2967	29.3615	25.8575	22.5516	19.2620	15.7635	11.5826	8.7270	4.7600
25	29.6000	23.1040	19.9136	16.3490	13.9994	12.1487	10.5533	9.0899	7.6708	6.2006	4.4914	3.3520	1.8020
50	15.7800	12.0620	10.2960	8.3646	7.1149	6.1424	5.3123	4.5574	3.8312	3.0848	2.2244	1.6552	.8870
100	8.1420	6.1620	5.2346	4.2305	3.5865	3.0883	2.6651	2.2819	1.9146	1.5387	1.1071	.8226	.4395
500	1.6705	1.2538	1.0609	.8539	.7220	.6204	.5345	.4569	.3828	.3072	.2206	.1637	.0874
1,000	.8380	.6282	.5314	.4274	.3613	.3104	.2673	.2285	.1914	.1536	.1103	.0818	.0436
5,000	.1690	.1259	.1064	.0856	.0723	.0622	.0535	.0458	.0383	.0307	.0220	.0164	.0087
10,000	.8404	.0629	.0532	.0428	.0362	.0311	.0267	.0229	.0191	.0154	.0110	.0082	.0044

$c = 3$

PROBABILITY OF ACCEPTANCE

SAMPLE SIZE	1%	5%	10%	20%	30%	40%	50%	60%	70%	80%	90%	95%	99%
10	70.2890	60.6630	55.1740	48.3658	43.4480	39.2996	35.5100	31.8397	28.0900	23.9450	18.7570	15.0030	9.3300
25	34.8790	28.1730	24.8019	20.9630	18.3900	16.3100	14.4925	12.7949	11.1137	9.3259	7.1670	5.6570	3.4480
50	18.7300	14.7838	12.8757	10.7543	9.3570	8.2525	7.2950	6.4096	5.5413	4.6280	3.5348	2.7788	1.6837
100	9.6980	7.5720	6.5586	5.4459	4.7204	4.1508	3.6598	3.2080	2.7671	2.3060	1.7559	1.3777	.8324
500	1.9950	1.5434	1.3313	1.1002	.9508	.8341	.7339	.6421	.5529	.4597	.3494	.2738	.1660
1,000	1.0010	.7736	.6669	.5508	.4759	.4173	.3671	.3211	.2765	.2298	.1746	.1368	.0825
5,000	.2008	.1550	.1336	.1103	.0952	.0835	.0734	.0642	.0553	.0459	.0349	.0273	.0165
10,000	.1004	.0775	.0668	.0551	.0476	.0418	.0367	.0321	.0276	.0230	.0175	.0137	.0082

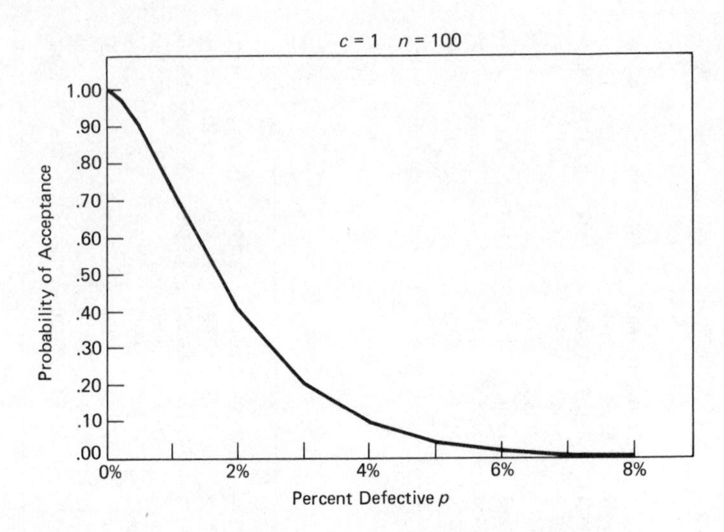

Figure 9.16. OC Curve generated from Table 9.4.

LTPD constant, these curves are especially useful. For example, let us say we are presented with a sampling plan calling for $n = 300$ units on test with $c = 3$ allowed failures. Reference to Figure 9.15 shows that this plan has an AQL of .45% and an LTPD of 2.25%. However, the product is very expensive and we would like to reduce the sample size while still retaining the LTPD (consumer's risk) at 2.25%. We look at Figure 9.14 for $c = 2$, find on the y axis the LTPD value of 2.25%, and go across the graph to the diagonal line labeled 10 (for probability of acceptance equal to 10%). From the intersection of this line with $p = 2.25\%$, we drop down to the x axis and read $n = 223$ units. Hence, an alternative sample plan to $n = 300$, $c = 3$ that has the same LTPD is $n = 223$ and $c = 2$.

In the same manner, by referring to Figures 9.13 and 9.12, we determine that the sampling plans having $n = 165$, $c = 1$, and $n = 100$, $c = 0$ preserve the LTPD of the original plan. So we can run the experiment with a minimum sample size of 100 units, allowing no failures, and be 90% confident that the percent defective of the population sampled is no higher than 2.25%. This last statement follows since subtracting the percent value on each diagonal line from 100% represents the one-sided confidence level at any given percent defective p value that the number of failures in the sample size n will be less than the acceptance number.

Note, however, by looking at the 95% diagonal line on each figure, that the AQL has gone from .45% at $n = 300$, $c = 3$ to .36%, .225%, and .053% for the $c = 2$, 1, and 0 plans, respectively. Thus, the product sampled must have a lower percent defective in order to consistently pass the plans having smaller acceptance numbers. Such is the price one pays to reduce the sample size while holding the same LTPD. A nontechnical way of showing this AQL shift is to observe that the allowed fallout to pass the test goes down from 3 failures out of 300 (or 1%), to 2 failures in 223 (or .9%), to 1 failure in 165 (or .6%), to zero failures in 100 (or 0%) as the sample size decreases. Of course, once a plan is chosen, we can immediately generate the OC curve to check the risks at other p values by the procedure described previously.

One final comment is made: if we calculate the ratio of the LTPD to the AQL for the $c = 0, 1, 2$, or 3 cases, we find the ratios equal to 43, 11, 6.5, and 4.9, respectively, independent of sample size. Thus, the larger the acceptance number, the smaller the difference between the AQL and LTPD; in other words, the OC curves become "steeper" as c increases.

MINIMUM SAMPLE SIZE PLANS

Suppose we wish to protect against a Type II error using the smallest size possible, and Type I error is not a primary concern. This situation may occur, for example, when parts are limited in availability or are highly expensive or time consuming to test.

Minimum sampling plans are based on an acceptance number of zero, that is, $c = 0$ since any number greater than zero would permit a larger sample size to be used for the same acceptable and rejectable (AQL and RQL) percent defective levels. If we assume the sample size is small or the sample is drawn from an ongoing process, then the binomial distribution applies. Hahn (1979) treats the situation where the lot size is a factor and develops the curves according to the hypergeometric distribution; at the lower percent defective values, however, considerable "eyeball" interpolation is involved. Here, we present curves for consumer protection levels of 90 and 95% for all percent defective values over the range of .001 (10 PPM) to 100% (1,000,000 PPM) based on the binomial distribution.

The minimum sample size graph is shown as Figure 9.17. As is indicated by the arrows, the right-hand pair of diagonal lines refer to the right-hand vertical axis; the left-hand pair refer to the left-hand vertical axis.

The derivation of the lines is quite simple. Since the acceptance number is zero, we want the probability of zero failures to be equal to the consumer risk level, that is, the β or Type II error. Thus, we have

$$(1 - p)^n = \beta.$$

Solving for n gives

$$n = \frac{\ln \beta}{\ln(1 - \beta)},$$

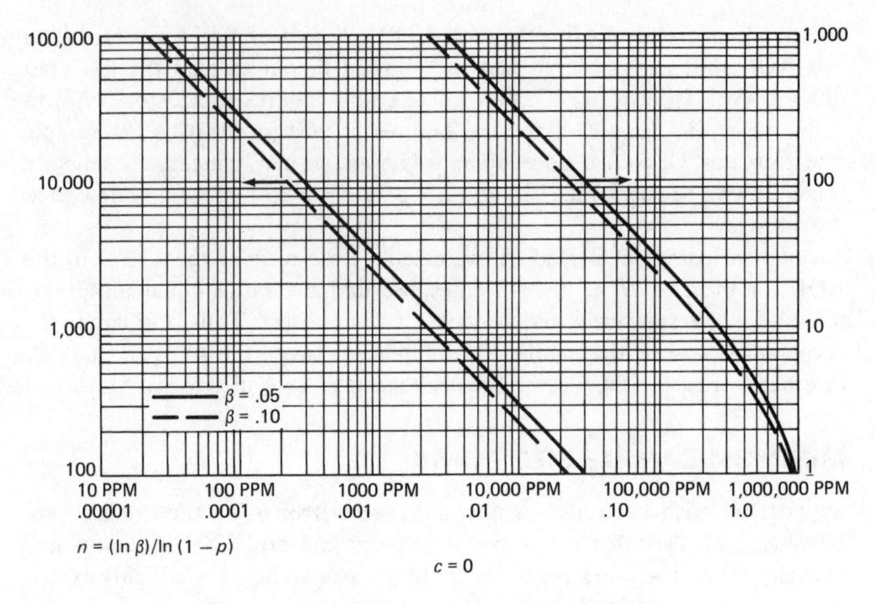

Figure 9.17. Graphs for minimum sampling plans.

as the minimum sample size necessary to assure a maximum Type II error risk of β if that population fraction defective is no higher than p. Figure 9.17 provides two curves, one for $\beta = .05$ (95% confidence) and the other for $\beta = .10$ (90% confidence or an LTPD plan). For example, to protect against a fraction defective higher than .015 (i.e., $p = 1.5\%$) with 90% confidence, the minimum sample size is about 150 units.

NEARLY MINIMUM SAMPLING PLANS

Some engineers (not to mention managers!) might feel uncomfortable about making a decision based on no failures. They feel there is some consideration that has to be given to the producer's risk, even at the cost of additional units. Thus, we also include a graph to find the sample size for nearly minimum sampling plans based on an acceptance number of $c = 1$. Figure 9.18 shows two consumer protection levels, $\beta = .05$ and $\beta = .10$. This type of plan allows at most one failure to occur before the lot or process is rejected. Of course, the sample sizes are higher than the minimum sampling case ($c = 0$) for the same confidence level and fraction defective.

The derivation of the nearly minimum sampling plan is straightforward.

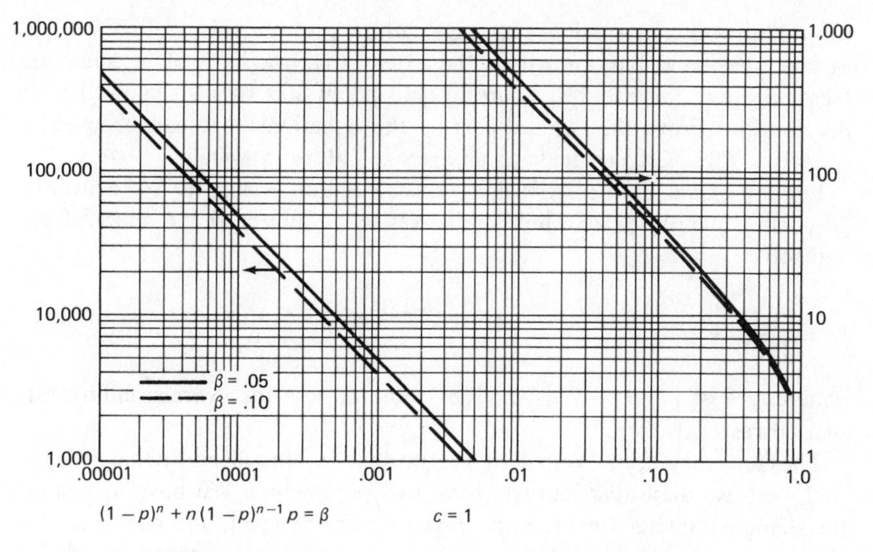

Figure 9.18. Graphs for nearly minimum sampling plans.

We want the probability of zero or 1 failure to be set at the consumer risk level. Then, based on the binomial distribution, we require

$$(1-p)^n + n(1-p)^{n-1}p = \beta.$$

This equation can not be solved explicitly for n, but it is a simple procedure to calculate corresponding sample sizes n for various p values and thereby generate the lines shown in Figure 9.18. As an example, one would need approximately 315 units to assure with 95% confidence that the population fraction defective is no higher than 1.5% (.015), assuming not more than 1 failure occurs in the sample.

RELATING AN OC CURVE TO LOT FAILURE RATES

Lot acceptance sampling plans can be designed to evaluate and assure reliability. A sample from a lot is tested for t hours at high stress and the proportion of reliability failures generated is the sample defect level. Intuitively, if this defect level is controlled and low enough, the product will have an acceptable failure rate. Often, however, it is difficult to quantify what defect level is acceptable and how any defect level relates to field failure rates.

In order to convert the proportion defective scale on an OC chart to a field failure rate scale we need a model that relates fallout at high stress for t hours to a normal use average failure rate. In this section we will assume that model is known to the extent that we have an acceleration factor A, that converts test hours to field hours and we also know the life distribution. From this we will derive the equations that can be used to transform the OC chart scale to an average failure rate scale.

For the exponential distribution, the calculation is simple. We convert a proportion p failures at t hours of stress to p failures at At hours of use and solve

$$1 - e^{-\lambda At} = p$$

obtaining $\lambda = [-\ln(1-p)]/At$. This value of λ is the average failure rate that corresponds to p.

The conversion for a Weibull or a lognormal requires an additional assumption, because these distributions have two parameters. We have to assume the shape parameter (m or σ) is known for the product, and does not vary significantly from lot to lot. Lot quality causes the characteristic life or T_{50} parameter to vary, accounting for changes in the failure rate. This is a strong assumption—but experience has shown it is often reasonable.

The equations to go from p to a Weibull AFR (assuming m is known) are

$$p = 1 - e^{-\left(\frac{At}{c}\right)^m}$$

$$c = \frac{At}{[-\ln(1-p)]^{1/m}}$$

$$\text{AFR}(U) = \frac{U^{m-1}[-\ln(1-p)]}{(At)^m},$$

where U is the period of use lifetime of interest for the average failure rate calculation. These equations come from the Weibull formulas in Chapter 4 and the average failure rate definition in Chapter 2.

The equations for a lognormal failure mode (assuming σ is known) are

$$p = \Phi\left[\frac{\ln (At/T_{50})}{\sigma}\right]$$

$$T_{50} = Ate^{-\sigma\Phi^{-1}(p)}$$

$$\text{AFR}(U) = \frac{-\ln[1 - \Phi((\ln U/T_{50})/\sigma)]}{\text{U}}.$$

These equations make use of the formulas for the lognormal (Chapter 5). The procedure is to use a p value in the second equation to calculate a corresponding T_{50} and then the third equation to obtain the equivalent AFR (Example 5.6 in Chapter 5 shows how to use the second equation to find T_{50}).

By using the formulas in this section, lot acceptance sampling plans can be defined in terms of an AQL field average failure rate and an LTPD field average failure rate.

STATISTICAL PROCESS CONTROL CHARTING FOR RELIABILITY

We have discussed procedures for lot acceptance. However, statistical process control techniques can also be used for monitoring the reliability of a process. Basically, the fraction (or percent) defective values from periodic samples are plotted on a graph. This chart, called a control chart, has a line, called an upper control limit (UCL), that indicates when the plotted fraction defective is significantly higher than that normally experienced. The difference between an UCL and an engineering specification is critical: the control limit is based on the previous history of the process; the engineering specification may have no relationship to the process capability. Thus, we can have a situation in which the process is "in spec" but yet out of control (and vice-versa)!

The UCL is simple to calculate. If p is the historical process average, then the UCL is given by

$$\mathrm{UCL} = p + z_{1-\alpha}\sqrt{\frac{p(1-p)}{n}},$$

where n is the periodic sample size and $z_{1-\alpha}$ is the standard normal variate value corresponding to a desired $(1-\alpha) \times 100\%$ confidence level. Typically $z_{1-\alpha}$ is set at 2 or 3 to roughly correspond to one-sided probabilities of 97.7 or 99.9%, respectively. The interpretation is that there are only 2 chances in 100 (or 1 chance in 1000) that a sample of size n would have a percent defective above the 2 (or 3) "sigma" limit. Therefore, when the control limit is exceeded, we assume the process is out of control and corrective action is required. Otherwise, the process is operating normally and no action is required. The term "sigma" refers to the fact that the equation for the UCL involves the term for the standard error of the binomial distribution along with the normal distribution approximation for the probabilities.

If the periodic sample size is nearly constant (that is, any one lot is within 30% of the average sample size), then it is not necessary to adjust the UCL

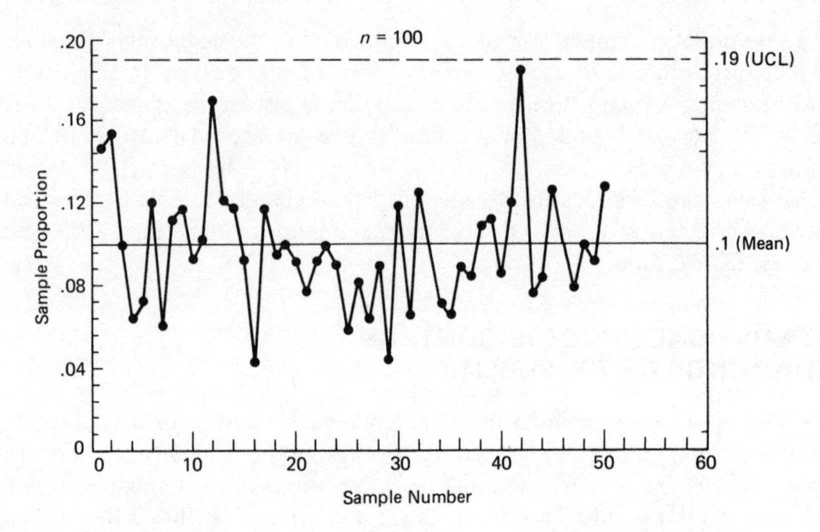

Figure 9.19. 3-sigma control chart for binomial proportions.

for each sample. Instead, one constant line will serve as the control limit. Figure 9.19 illustrates a typical control chart for attribute data.

There are also control charts for variables data involving both the sample means and the sample ranges and for Poisson distributed data. Many excellent books on the subject of statistical quality control exist: see Burr (1976,1979), Deming (1982), Duncan (1965), Grant and Leavenworth (1980), Ishikawa (1982), Juran (1979), Ott (1975), or the *Western Electric Handbook* (1958). The interested reader involved in monitoring and improving process reliability should consult these references.

SUMMARY

In this chapter, we considered the relationship of quality control concepts to reliability problems. Starting with simple ideas on permutations and combinations, we developed the binomial distribution. We talked about how binomial estimates could be used to nonparametrically provide probability information in the absence of knowledge about the underlying distribution of fail times. We showed several methods for calculating binomial confidence intervals. We also discussed two other discrete distributions: the hypergeometric and the Poisson. We showed the correspondence between the discrete Poisson and the continuous exponential distributions. We discussed various types of sampling, quality concepts such as AQL and LTPD, associated risks, properties of operating characteristic curves, and the selection of sample sizes. We illustrated many procedures, graphical and otherwise, to simplify the generation, implementation, and comparison of various sampling plans. The application of minimum and near-minimum sampling plans was covered. The relation of failure rates to sampling considerations was highlighted. Finally, we showed how statistical process control could be applied to the monitoring of reliability situations.

Bibliography

Abramovitz, M. and Stegun, I. A., Ed. (1964), *Handbook of Mathematical Functions,* National Bureau of Standards, Washington, D.C.

Aitchison, J. and Brown, J. A. C. (1957), *The Log-normal Distribution,* Cambridge University Press, New York.

Ament, R. (1977), "Improved chart of confidence limits for p in binomial sampling," *PAS Reporter,* 15(10).

Barlow, R. E. and Proschan, F. (1975), *Statistical Theory of Reliability and Life Testing,* Holt, Rinehart and Winston, New York.

Burr, I. W. (1976), *Statistical Quality Control Methods,* Marcel Dekker, New York.

Burr, I. W. (1979), *Elementary Statistical Quality Control,* Marcel Dekker, New York.

Chace, E. F. (1976), "Right-censored grouped life test data analysis assuming a two parameter Weibull distribution function," *Microelectronics and Reliability,* 15:497–499.

Chemical Rubber Company (CRC) (1967), *Standard Mathematical Tables.* 15th ed., Cleveland.

Clopper, C. J. and Pearson, E. S. (1934), "The use of confidence or fiducial limits illustrated in the case of the binomial," *Biometrika,* 26:404.

Cohen, A. C., Jr. (1959), "Simplified estimators for the normal distribution when samples are singly censored or truncated," *Technometrics* 1:217–237.

Deming, W. E. (1982), *Quality, Productivity, and Competitive Position,* Massachusetts Institute of Technology, Cambridge.

Dixon, W. J. and Massey, F. J., Jr. (1969), *Introduction to Statistical Analysis,* 3rd ed., McGraw-Hill, New York.

Dodge, H. F. and Romig, H. G. (1959), *Sampling Inspection Tables, Single and Double Sampling,* 2nd ed., John Wiley and Sons, New York.

Duke, S. D. and Meeker, W. Q., Jr. (1981), "CENSOR—a user oriented computer program for life data analysis," *The American Statistician,* 34:59–60.

Duncan, A. J. (1965), *Quality Control and Industrial Statistics,* 3rd ed., Richard D. Irwin, Homewood, IL.

Eyring, H., Glasstones, S. and Laidler, K. J. (1941), *The Theory of Rate Processes,* McGraw-Hill, New York.

Feller, W. (1968), *An Introduction to Probability Theory and Its Applications,* Vol. 1: 3rd ed., Wiley, New York.

Gnedenko, B. V., Belyayev, Y. K. and Solovyev, A. D. (1969), *Mathematical Methods of Reliability Theory,* Academic Press, New York.

Grant, E. L. and Leavenworth, R. S. (1980), *Statistical Quality Control,* 5th ed., McGraw-Hill, New York.

Gumbel, E. J. (1954), *Statistical Theory of Extreme Values and Some Practical Applications,* National Bureau of Standards, Washington, D.C.

Guenther, W. C. (1974), "Sample size formulas for some binomial problems," *Technometrics,* 16:465–467.

Hahn, G. J. (1979), "Minimum and near minimum sampling plans," *Journal of Quality Technology,* 11(4):206–212.

Hahn, G. J. and Shapiro, S. S. 1967), *Statistical Models in Engineering,* Wiley, New York.

Ishikawa, K. (1982), *Guide to Quality Control,* Asian Productivity Organization, Tokyo.

Jensen, F. and Petersen, N. E. (1982), *Burn-in: an Engineering Approach to the Design and Analysis of Burn-in Procedures,* Wiley, New York.

Johnson, L. G. (1951), "The median ranks of sample values in their population with an application to certain fatigue studies," *Industrial Mathematics,* 2:1–9.

Juran, J. M., ed. (1974), *Quality Control Handbook,* 3rd ed., McGraw-Hill, New York.

Kaplan, E. L. and Meier, P. (1958), "Nonparametric Estimation from Incomplete Observations," *Journal of the American Statistical Association,* 53:457–481.

Kielpinski, T. J. and Nelson, W. (1975), "Optimum censored accelerated life-tests for the normal and lognormal life distributions," *IEEE Transactions on Reliability,* R-24(5):310–320.

Kolmogorov, A. N. (1941), "On a logarithmic normal distribution law of the dimensions of particles under pulverization," *Dokl. Akad. Nauk, USSR,* 31(2):99–101.

Landzberg, A. H. and Norris, K. C. (1969), "Reliability of controlled collapse interconnections," *IBM Journal of Research and Development,* 13(3).

Larson, H. R. (1966), "A nomograph of the cumulative binomial distribution," *Industrial Quality Control,* 23(6):270–278.

Lawless, J. F. (1982), *Statistical Models and Methods for Lifetime Data,* Wiley, New York.

Mann, N. R. Schafer, R. E. and Singpurwalla, N. D. (1974), *Methods for Statistical Analysis of Reliability and Life Data,* Wiley, New York.

Meeker, W. Q. and Nelson, W. (1975), "Optimum accelerated life-tests for the Weibull and extreme value distributions," *IEEE Transactions on Reliability,* R-24(5):321–332.

Mendenhall, W. and Scheaffer, R. L. (1973), *Mathematical Statistics with Applications,* Wadsworth, Belmont, CA.

Nelson, W. (1969), "Hazard plotting for incomplete failure data," *Journal of Quality Technology,* 1:27–52.

Nelson, W. (1972), "Theory and applications of hazard plotting for censored failure data," *Technometrics,* 14:945–966.

Nelson, W. (1975), "Graphical analysis of accelerated life test data with a mix of failure modes," *IEEE Transactions on Reliability,* R-24(4):230–237.

Nelson, W. (1982), *Applied Life Data Analysis,* Wiley, New York.

Ostle, B. and Mensing, R. W. (1975), *Statistics in Research,* 3rd ed., The Iowa State University Press, Ames, Iowa.

Ott, E., (1975), *Process Quality control,* McGraw-Hill, New York.

Ott, L. (1977), *An Introduction to Statistical Methods and Data Analysis,* Wadsworth, Belmont, CA.

Peck, D. and Trapp, O. D. (1980), *Accelerated Testing Handbook,* Technology Associates and Bell Telephone Laboratories, Portola, CA.

Prins, J. (1984), STATLIB: Statistical Software for the IBM Personal Computer, IBM Personally Developed Software, P.O. Box 3280, Wallingford, CT.

Schilling, E. G. (1982), *Acceptance Sampling in Quality Control,* Marcel Dekker, New York.

Strauss, S. H. (1980), "STATPAC: a general purpose package for data analysis and fitting statistical models to data," *The American Statistician,* 34:59–60.

Thomas, G. B., Jr. (1960), *Calculus and Analytical Geometry,* 3rd ed., Addison-Wesley, Reading, MA.

Trindade, D. C. (1980), "Nonparametric Estimation of a Lifetime Distribution via the Renewal Function," Ph.D. Dissertation, University of Vermont, Burlington, VT.

Trindade, D. C. and Haugh, L. D. (1979), "Nonparametric Estimation of a Lifetime Distribution via the Renewal Function," IBM Technical Report, TR 19.0463, Burlington, VT.

Trindade, D. C. and Haugh, L. D. (1980), "Estimation of the reliability of computer components from field renewal data," *Microelectronics and Reliability,* 20:205–218.

United States Department of Defense (1963), Military Standard, *Sampling Procedures and Tables for Inspection by Attributes* (MIL-STD-105D), U.S. Government Printing Office.

United States Department of Defense (1963), Military Specification, *General Specification for Semiconductor Devices* (MIL-S-19500G), U.S. Government Printing Office.

Weibull, W. (1951), "A statistical distribution function of wide applicability," *Journal of Applied Mechanics,* 18:293–297.

Western Electric Co. (1958), *Statistical Quality Control Handbook,* 2nd ed., Mack, Easton, PA.

Wilks, S. S. (1962), *Mathematical Statistics,* Wiley, New York.

Appendix
Percentiles of the χ^2 Distribution[a]

v \ q	0.005	0.010	0.025	0.05	0.10	0.20	0.30	0.40
1	0.0^4393	0.0^3157	0.0^3982	0.0^2393	0.0158	0.0642	0.148	0.275
2	0.0100	0.0201	0.0506	0.103	0.211	0.446	0.713	1.02
3	0.0717	0.115	0.216	0.352	0.584	1.00	1.42	1.87
4	0.207	0.297	0.484	0.711	1.06	1.65	2.19	2.75
5	0.412	0.554	0.831	1.15	1.61	2.34	3.00	3.66
6	0.676	0.872	1.24	1.64	2.20	3.07	3.83	4.57
7	0.989	1.24	1.69	2.17	2.83	3.82	4.67	5.49
8	1.34	1.65	2.18	2.73	3.49	4.59	5.53	6.42
9	1.73	2.09	2.70	3.33	4.17	5.38	6.39	7.36
10	2.16	2.56	3.25	3.94	4.87	6.18	7.27	8.30
11	2.60	3.05	3.82	4.57	5.58	6.99	8.15	9.24
12	3.07	3.57	4.40	5.23	6.30	7.81	9.03	10.2
13	3.57	4.11	5.01	5.89	7.04	8.63	9.93	11.1
14	4.07	4.66	5.63	6.57	7.79	9.47	10.8	12.1
15	4.60	5.23	6.26	7.26	8.55	10.3	11.7	13.0
16	5.14	5.81	6.91	7.96	9.31	11.2	12.6	14.0
17	5.70	6.41	7.56	8.67	10.1	12.0	13.5	14.9
18	6.26	7.01	8.23	9.39	10.9	12.9	14.4	15.9
19	6.84	7.63	8.91	10.1	11.7	13.7	15.4	16.9
20	7.43	8.26	9.59	10.9	12.4	14.6	16.3	17.8
21	8.03	8.90	10.3	11.6	13.2	15.4	17.2	18.8
22	8.64	9.54	11.0	12.3	14.0	16.3	18.1	19.7
23	9.26	10.2	11.7	13.1	14.8	17.2	19.0	20.7
24	9.89	10.9	12.4	13.8	15.7	18.1	19.9	21.7
25	10.5	11.5	13.1	14.6	16.5	18.9	20.9	22.6
26	11.2	12.2	13.8	15.4	17.3	19.8	21.8	23.6
27	11.8	12.9	14.6	16.2	18.1	20.7	22.7	24.5
28	12.5	13.6	15.3	16.9	18.9	21.6	23.6	25.5
29	13.1	14.3	16.0	17.7	19.8	22.5	24.6	26.5
30	13.8	15.0	16.8	18.5	20.6	23.4	25.5	27.4
35	17.2	18.5	20.6	22.5	24.8	27.8	30.2	32.3
40	20.7	22.2	24.4	26.5	29.1	32.3	34.9	37.1
45	24.3	25.9	28.4	30.6	33.4	36.9	39.6	42.0
50	28.0	29.7	32.4	34.8	37.7	41.4	44.3	46.9
75	47.2	49.5	52.9	56.1	59.8	64.5	68.1	71.3
100	67.3	70.1	74.2	77.9	82.4	87.9	92.1	95.8

Percentiles of the χ^2 Distribution (*continued*)[a]

ν \ q	0.50	0.60	0.70	0.80	0.90	0.95	0.975	0.990	0.995	0.999
1	0.455	0.708	1.07	1.64	2.71	3.84	5.02	6.63	7.88	10.8
2	1.39	1.83	2.41	3.22	4.61	5.99	7.38	9.21	10.6	13.8
3	2.37	2.95	3.67	4.64	6.25	7.81	9.35	11.3	12.8	16.3
4	3.36	4.04	4.88	5.99	7.78	9.49	11.1	13.3	14.9	18.5
5	4.35	5.13	6.06	7.29	9.24	11.1	12.8	15.1	16.7	20.5
6	5.35	6.21	7.23	8.56	10.6	12.6	14.4	16.8	18.5	22.5
7	6.35	7.28	8.38	9.80	12.0	14.1	16.0	18.5	20.3	24.3
8	7.34	8.35	9.52	11.0	13.4	15.5	17.5	20.1	22.0	26.1
9	8.34	9.41	10.7	12.2	14.7	16.9	19.0	21.7	23.6	27.9
10	9.34	10.5	11.8	13.4	16.0	18.3	20.5	23.2	25.2	29.6
11	10.3	11.5	12.9	14.6	17.3	19.7	21.9	24.7	26.8	31.3
12	11.3	12.6	14.0	15.8	18.5	21.0	23.3	26.2	28.3	32.9
13	12.3	13.6	15.1	17.0	19.8	22.4	24.7	27.7	29.8	34.5
14	13.3	14.7	16.2	18.2	21.1	23.7	26.1	29.1	31.3	36.1
15	14.3	15.7	17.3	19.3	22.3	25.0	27.5	30.6	32.8	37.7
16	15.3	16.8	18.4	20.5	23.5	26.3	28.8	32.0	34.3	39.3
17	16.3	17.8	19.5	21.6	24.8	27.6	30.2	33.4	35.7	40.8
18	17.3	18.9	20.6	22.8	26.0	28.9	31.5	34.8	37.2	42.3
19	18.3	19.9	21.7	23.9	27.2	30.1	32.9	36.2	38.6	43.8
20	19.3	21.0	22.8	25.0	28.4	31.4	34.2	37.6	40.0	45.3
21	20.3	22.0	23.9	26.9	29.6	32.7	35.5	38.9	41.4	46.8
22	21.3	23.0	24.9	27.3	30.8	33.9	36.8	40.3	42.8	48.3
23	22.3	24.1	26.0	28.4	32.0	35.2	38.1	41.6	44.2	49.7
24	23.3	25.1	27.1	29.6	33.2	36.4	39.4	43.0	45.6	51.2
25	24.3	26.1	28.2	30.7	34.4	37.7	40.6	44.3	46.9	52.6
26	25.3	27.2	29.2	31.8	35.6	38.9	41.9	45.6	48.3	54.1
27	26.3	28.2	30.3	32.9	36.7	40.1	43.2	47.0	49.6	55.5
28	27.3	29.2	31.4	34.0	37.9	41.3	44.5	48.3	51.0	56.9
29	28.3	30.3	32.5	35.1	39.1	42.6	45.7	49.6	52.3	58.3
30	29.3	31.3	33.5	36.3	40.3	43.8	47.0	50.9	53.7	59.7
35	34.3	36.5	38.9	41.8	46.1	49.8	53.2	57.3	60.3	66.6
40	39.3	41.6	44.2	47.3	51.8	55.8	59.3	63.7	66.8	73.4
45	44.3	46.8	49.5	52.7	57.5	61.7	65.4	70.0	73.2	80.1
50	49.3	51.9	54.7	58.2	63.2	67.5	71.4	76.2	79.5	86.7
75	74.3	77.5	80.9	85.1	91.1	96.2	100.8	106.4	110.3	118.6
100	99.3	102.9	106.9	111.7	118.5	124.3	129.6	135.6	140.2	149.4

[a] Abridged from Table V of A. Hald, "Statistical Tables and Formulas," 1952, Wiley, New York.

Index

Index